SEA-LEVEL CHANGE IN THE PACIFIC ISLANDS REGION

A REVIEW OF EVIDENCE TO INFORM ASIAN DEVELOPMENT BANK GUIDANCE ON SELECTING SEA-LEVEL PROJECTIONS FOR CLIMATE RISK AND ADAPTATION ASSESSMENTS

JULY 2022

ASIAN DEVELOPMENT BANK

© 2022 Asian Development Bank
6 ADB Avenue, Mandaluyong City, 1550 Metro Manila, Philippines
Tel +63 2 8632 4444; Fax +63 2 8636 2444
www.adb.org

Some rights reserved. Published in 2022.

ISBN 978-92-9269-644-3 (print); 978-92-9269-645-0 (electronic); 978-92-9269-646-7 (ebook)
Publication Stock No. TCS220312-2
DOI: http://dx.doi.org/10.22617/TCS220312-2

How to cite this publication: ADB. 2022. *Sea-Level Change in the Pacific Islands Region: A Review of Evidence to Inform Asian Development Guidance on Selecting Sea-Level Projections for Climate Risk and Adaptation Assessments*. Manila. DOI: http://dx.doi.org/10.22617/TCS220312-2

Notes:
In this publication, "$" refers to United States dollars.
ADB recognizes "United States of America" and "USA" as the United States.

Cover design by Alfredo De Jesus. On the cover: Aerial shot of Aitutaki Island, one of the most visited places in the Cook Islands (photo by ADB).

 Printed on recycled paper

Contents

Tables and Figures

Acknowledgments

This report was prepared by Anthony Kiem, a hydroclimatologist from the Centre for Water, Climate and Land (CWCL) at the University of Newcastle, Australia (Contact: Anthony.Kiem@newcastle.edu.au).

Experts working on different aspects of sea-level rise (SLR) discussed and reviewed this report, including in two virtual workshops:

(i)   The first was held on 20 October 2020 to discuss the first draft, identify any literature that had been missed but should be considered, and identify and resolve any parts of the report that were unclear or incorrect.

(ii) The second was held on 24 November 2020 to confirm that the updated report accurately summarized the relevant science and to discuss and develop the outline for the guidance on how to incorporate credible SLR information into Asian Development Bank (ADB) projects.

The participants in these workshops are listed below. The significant time and input they provided to review the evidence is greatly appreciated.

Andra J. Garner	Rowan University, USA
Andrew Magee	Centre for Water, Climate and Land (CWCL), University of Newcastle, Australia
Antoine N'Yeurt	University of South Pacific, Suva, Fiji
Charles Rodgers	Sustainable Development and Climate Change Department, ADB
Chip Fletcher	University of Hawaii, USA
Connon Andrews	Market Director for Climate Resilience and Adaptation, Beca Group, New Zealand
Curt Storlazzi	United States Geological Survey, University of California Santa Cruz, USA
Dennis Fenton	ADB Consultant, Support to Climate Resilient Investment Pathways in the Pacific
Elisabeth Holland	University of South Pacific, Suva, Fiji
Herve Damlanian	The Pacific Community, New Caledonia
Jeff Bowyer	ADB Consultant, Support to Climate Resilient Investment Pathways in the Pacific

Jimaima Lako	Fiji National University, Suva, Fiji
Jonathan L. Bamber	University of Bristol, United Kingdom
Kathy McInnes	Commonwealth Scientific and Industrial Research Organisation, Australia
Kevin Walsh	University of Melbourne, Australia
Lovelyn Otoiasi	Solomon Islands National University
Mark Hemer	Commonwealth Scientific and Industrial Research Organisation, Australia
Mike Beck	University of California, Santa Cruz, USA
Molly Powers	The Pacific Community, New Caledonia
Morgan Wairiu	University of South Pacific, Suva, Fiji
Moritz Wandres	The Pacific Community, New Caledonia
Noelle O'Brien	Sustainable Development and Climate Change Department, ADB
Paul Watkiss	Paul Watkiss Associates, United Kingdom
Rob Wilby	Loughborough University, United Kingdom
Robert E. Kopp	Rutgers University, USA
Robert Nicholls	University of East Anglia, United Kingdom

Abbreviations

ADB   –   Asian Development Bank
AR5   –   Fifth Assessment Report of the IPCC
AR6   –   Sixth Assessment Report of the IPCC
CRA   –   climate risk and adaptation assessment
CSIRO   –   Commonwealth Scientific and Industrial Research Organisation
DEM   –   digital elevation model
ENSO   –   El Niño–Southern Oscillation
IPCC   –   Intergovernmental Panel on Climate Change
IPO   –   Interdecadal Pacific Oscillation
m   –   meter
mm   –   millimeter
NASA   –   United States National Aeronautics and Space Administration
PACCSAP   –   Pacific-Australia Climate Change Science and Adaptation Planning
PARD   –   ADB Pacific Department
PIR   –   Pacific Islands Region
RCP   –   Representative Concentration Pathway
SLR   –   sea-level rise
SROCC   –   Special Report on the Oceans and Cryosphere in a Changing Climate
SSP   –   Shared Socioeconomic Pathway
SST   –   sea surface temperature

Executive Summary

The engagement in the Pacific Islands Region (PIR) of the Pacific Department (PARD) of the Asian Development Bank (ADB) (Figure 1) includes supporting developing countries in the PIR through (i) regional development forums and infrastructure finance, (ii) regional projects focused on renewable energy and marine and coastal management, and (iii) strengthening of disaster preparedness. PARD's regional technical assistance also includes contributing to developing capacity for public financial management, statistics, and data collection.

The PIR, especially the western tropical Pacific, is particularly vulnerable to sea-level rise (SLR) because of (i) high exposure to tropical cyclones and other tropical storms; (ii) high shoreline to land area ratios; (iii) high sensitivity to changes in sea level, waves, and currents; and (iv) its many low-lying coral atolls, reefs, or volcanically composed islands.

Given the vulnerability of the PIR to SLR, how precautionary should investors be when dealing with SLR in the Pacific? Which source, or combination of sources, of SLR projection information should investors use in climate risk and adaptation assessments (CRAs) and what should be considered when investing in the PIR? To answer these questions, this report reviews the evidence to establish which sources of SLR projections are credible for the PIR, as well as the strengths, weaknesses, and uncertainties associated with various sources of information on SLR in the PIR. This review of evidence is intended to lay out advisory standards for planning, testing, and design of investments in the PIR. This report focuses on reviewing SLR science and evidence to provide credible information about SLR in the PIR that can then be used as the basis for further investigation into, and decision-making on, bigger picture issues such as long-term planning for sustainable settlements and any resulting national/regional population movements.

Projections from the Fifth Assessment Report of the Intergovernmental Panel on Climate Change (AR5), published in 2013, suggested that SLR in the PIR is unlikely to exceed 1 meter (m) by 2100 (relative to the 1986–2005 baseline used in AR5). This information has been widely used by ADB and others to assess and manage SLR-related risks in the PIR, with allowance and adaptation for SLR of up to 1 m considered sufficiently precautionary for projects with operational lifetimes of less than 100 years. However, the key finding from this review of science and evidence that has emerged since the publication of AR5 is that this approach may not always be adequate in the PIR for the following reasons:

(i)   Although there is very high confidence in the direction of change for all PIR locations (i.e., a fall in sea level is not projected anywhere in the PIR), there is only medium confidence in the magnitude of change. Nevertheless, for most locations in the PIR where location-specific analysis has been conducted and where the impacts of natural climate variability are considered, it is possible that SLR might exceed 1 m by the end of the 21st century (relative to the 1986–2005 baseline used in AR5).

(ii)   The Sixth Assessment Report of the Intergovernmental Panel on Climate Change (AR6) and other work that has emerged since AR5 demonstrate that not only is SLR greater than 1 m (relative to the 1995–2014 baseline) conceivable at some point in the 21st century but it is also plausible that SLR could exceed 2 m by 2100. It is also important to note that SLR will not stop in 2100.

(iii)   Some paleoclimate records suggest that SLR of 5 m in a century has occurred before. However, the consensus view is that such extreme SLR would happen over long periods (centuries to millennia) and is unlikely to occur before 2100. AR6 states that projected global mean SLR of 1.7–6.8 m by 2300 is possible without marine ice cliff instability and this projection increases to 16 m by 2300 with marine ice cliff instability.

(iv)   Short-term variability in high-water levels associated with storm surge and waves could significantly increase local coastal water levels above what is expected because of changes to absolute sea level (i.e., changes to long-term average sea level alone), especially in the PIR and particularly in the western tropical Pacific (see previous page, second paragraph).

(v)   Based on observed data collected since ~2000, most islands in the PIR are subsiding (i.e., have negative vertical land movement). Therefore, irrespective of any other influence, the effect of SLR will be magnified where the land is falling, and this appears to be the case for much of the PIR.

While recognizing all the factors that contribute to SLR impacts (e.g., SLR, storm surge, waves, and vertical land movements, etc.), the findings of this review highlight the need for a more precautionary approach than simply adopting the upper global mean SLR projections when considering the impacts of high-water levels in CRAs for the PIR. There is a need to consider higher-end scenarios and to recognize that SLR will not stop in 2100. Using 2100 as a planning time frame is arbitrary and using a 100-year planning time frame (i.e., 2122) may be more appropriate for long-term planning. Therefore, it is advised that a precautionary approach for ADB CRAs in the PIR requires the following to be considered relative to the 1995–2014 baseline introduced in AR6:

(i)   for all projects, a 1 m SLR scenario, for comparison with existing studies that have typically used a scenario of 1 m SLR by 2100;

(ii)   for short- to medium-term projects (i.e., with a design life of 20–30 years), a scenario of 0.5 m SLR by 2050;

(iii)   for long-term projects (i.e., with a design life greater than 30 years), a scenario of 2 m SLR by 2100; and

(iv)   for projects with an expected lifetime beyond 2100, scenarios of greater than 2 m SLR.

These SLR scenarios should be used not only in sensitivity analyses of climate proofing design and options but also in the analysis of the costs and benefits of additional climate proofing, to explore the flexibility of adaptation options (and to avoid maladaptation). It is emphasized that these scenarios are recommended for sensitivity analysis rather than as minimum precautionary levels for climate proofing. The flexibility provided by adaptive management approaches could also address higher SLR, noting this needs to consider the lifetime, risk of lock-in, and level of precaution associated with investments. Where warranted (i.e., at sites with high exposure and/or vulnerability), extra allowance should also be made for the influence of natural climate variability, tropical cyclones, storm surge, waves, and vertical land movements. Exactly what that allowance should be will depend on the type of project and its location in the PIR, as well as the appetite for risk and expected lifetime of the project. Path dependency should also be considered. That is, decisions to invest (or otherwise) in coastal infrastructure based on assumptions about possible high-water levels (and other factors, including economics) will influence subsequent investments and development, for which the same level of risk assessment might not be performed. Refer to Pacific Region Infrastructure Facility (PRIF) (2021) for further guidance.

Recommendations are provided for future work required to develop and incorporate credible SLR projection information into ADB CRAs in the PIR. These include defining the objectives, tasks, estimated time for each task, and skills and personnel required to complete the tasks. The approach to assess and deal with SLR in the PIR could be linked to the SLR calculator and the associated knowledge product, which ADB has already developed for Viet Nam, to ensure transparency and consistency in the approach across ADB.

The science on SLR is evolving, and views are changing about how best to deal with existing and projected impacts of SLR. Adaptation to the impacts of climate change (including SLR) is also an ongoing process. It is recommended that the evidence should be reviewed every 5–10 years (ideally as soon as possible after the release of each new Intergovernmental Panel on Climate Change Assessment Report) to determine whether advice has changed in light of new evidence and, if required, to update the recommendations and guidance so that ADB activities are consistent with the best science and practice. The suitability and robustness of SLR adaptation strategies should also be reviewed every 5–10 years either to confirm that they are appropriate or, if needed, to implement other actions on the adaptation pathway.

I. Introduction

A. Project Description

The Asian Development Bank (ADB) Pacific Department's (PARD's) engagement in the Pacific Islands Region (PIR) (Figure 1) includes supporting developing countries in the PIR through (i) regional development forums and infrastructure finance, (ii) regional projects focused on renewable energy and marine and coastal management, and (iii) strengthening of disaster preparedness. PARD's regional technical assistance also includes contributing to developing capacity for public financial management, statistics, and data collection.

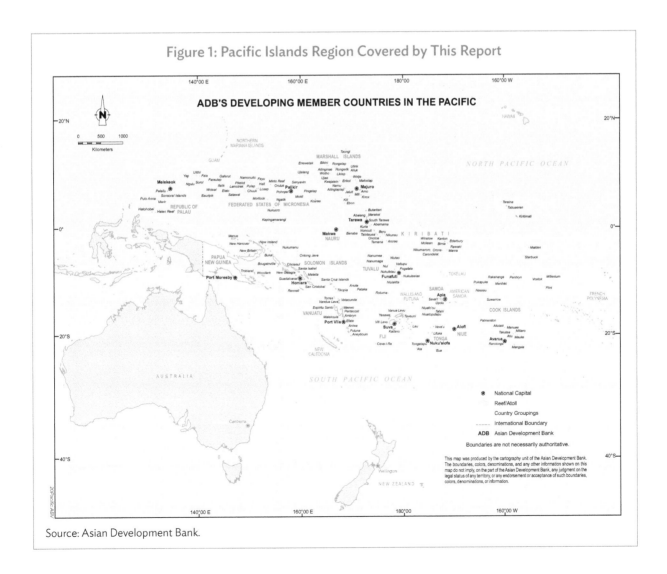

Figure 1: Pacific Islands Region Covered by This Report

Source: Asian Development Bank.

The PIR (Figure 1), especially the western tropical Pacific, is particularly vulnerable to sea-level rise (SLR) (Church et al. 2006) because of (i) high shoreline to land area ratios (Barnett 2001); (ii) high sensitivity to changes in coastal sea level, waves, and currents (Becker et al. 2012); and (iii) its many low-lying coral atolls, reefs, or volcanically composed islands (Connell 2013; Quataert et al. 2015; Vitousek et al. 2017).

Given the vulnerability of the PIR to SLR, how precautionary should investors be when dealing with SLR in the Pacific? Which source, or combination of sources of SLR projection information should investors use in climate risk and adaptation assessments (CRAs) and what should be considered when investing in the PIR? To answer these questions, this report reviews the evidence to establish which sources of SLR projections are credible for the PIR, as well as the strengths, weaknesses, and uncertainties associated with various sources of information on SLR in the PIR. This review of evidence is intended to lay out advisory standards for planning, testing, and design of investments in the PIR.

Once credible sources of SLR projection information are identified, advice based on the review of the latest science and good practice is required to assist with the development of PIR-specific guidance on how such information should be incorporated into projects (e.g., at the project feasibility and CRA stages). This process has been followed for other regions in the ADB portfolio (e.g., Viet Nam) (ADB 2020a, 2020b), and the approach to assessing and dealing with SLR in the PIR should be linked to this existing work to ensure consistency in approach across ADB. This report focuses on reviewing SLR science and evidence to provide credible information about SLR in the PIR that can then be used as the basis for further investigation into, and decision-making on, bigger picture issues such as long-term planning for sustainable settlements and any resulting national/regional population movements.

B.   Objectives and Scope of This Report

The objectives of this report are as follows:

(i)   Summarize the state of understanding of the causes and impacts of SLR across the PIR, how sea levels across the PIR have changed in the past, and how sea levels across the PIR are projected to change in the future.
(ii)   Identify what sea-level change data (historical) and projections (future) exist for the PIR and evaluate their relative strengths and weaknesses.
(iii)   Document the major uncertainties and scientific challenges that exist in relation to understanding and quantifying past and future sea-level change in the PIR.
(iv)   Provide advice, based on the latest science and good practice, to assist PARD in developing PIR-specific guidance on how to incorporate credible SLR projection information into ADB projects (e.g., at the project feasibility and CRA stages).
(v)   Recommend what more needs to be done to develop and implement guidance on incorporating credible SLR projections into ADB projects, including defining objectives, tasks, estimated time for each task, and skills and personnel required to complete the tasks.[1]

---

[1]   Includes recommendations on what tasks should be done as desktop studies, which should be tackled via meetings and workshops involving experts (key people that should be invited will be identified), and which require field work or data collection.

II. Understanding Sea-Level Change in the Pacific Islands Region

A. Natural Variability of Sea Level in the Pacific Islands Region

Sea-level changes can occur because of isolated, extreme events (e.g., storm surge) but more commonly arise from a combination of natural phenomena that individually may not be extreme (Nicholls et al. 2014, 2021; McInnes et al. 2016b). These natural phenomena occur over a range of time and space scales in any given PIR coastal location, and thus the contribution of each phenomenon to extreme sea levels varies (Figure 2). For example:

(i) Sea-level variability is high across the PIR, where monthly, seasonal, and interannual sea-level anomalies are highly correlated with ocean-atmospheric modes such as the El Niño–Southern Oscillation (ENSO) and the Interdecadal Pacific Oscillation (IPO) (e.g., Oliver and Thompson 2010; Becker et al. 2012; Zhang and Church 2012; White et al. 2014). See section II-B for further details.

(ii) Astronomical tides vary over multiple timescales (e.g., diurnal and semidiurnal, fortnightly with spring and neap tides, and on seasonal to interannual timescales) and are often the largest contributor to both sea-level variability and annual maximum sea-level elevation (relative to mean sea level) in the PIR (Stephens et al. 2014), especially for PIR locations that are not influenced by tropical cyclone activity (Merrifield et al. 2007). Astronomical tides also vary in space, with the highest spring tides across the PIR ranging from 0.6 m in French Polynesia and the Cook Islands to ~2 m in the eastern parts of the Federated States of Micronesia (Ramsay 2011).

(iii) Storm surges are gravity waves arising from the inverse barometric effect and wind stress (Walsh et al. 2012; McInnes et al. 2016b). The inverse barometric effect elevates sea levels by about 1 centimeter (cm) for every 1 hectopascal drop in atmospheric pressure relative to surrounding conditions. Wind stress refers to winds blowing from the ocean to land, causing an increase in sea levels (i.e., wind setup), particularly within semi-enclosed bays, and/or under severe wind such as that produced by tropical cyclones. The magnitude of storm surge is also determined by storm track, storm intensity, bathymetry, and the shape of the coastline.

(iv) Wave setup is the increase in water level landward of the breaking point of waves, and wave run-up is the additional height reached by a wave on a beach before its energy is dissipated by gravity and friction (Hoeke et al. 2013; McInnes et al. 2016b; Wandres et al. 2020). The magnitude of wave setup and run-up increases with the breaking height of the wave and when sea level is elevated because of SLR or seasonal to interannual variability, tides, and/or storm surge. Because wave setup and run-up are caused by breaking waves, sheltered coastal areas such as harbors and lagoons generally do not experience these effects. This is important to note because sheltered coastal areas are the typical location for tide gauges, meaning the impacts of wave setup and run-up on sea-level changes are typically not captured or are underestimated by tide-gauge data.

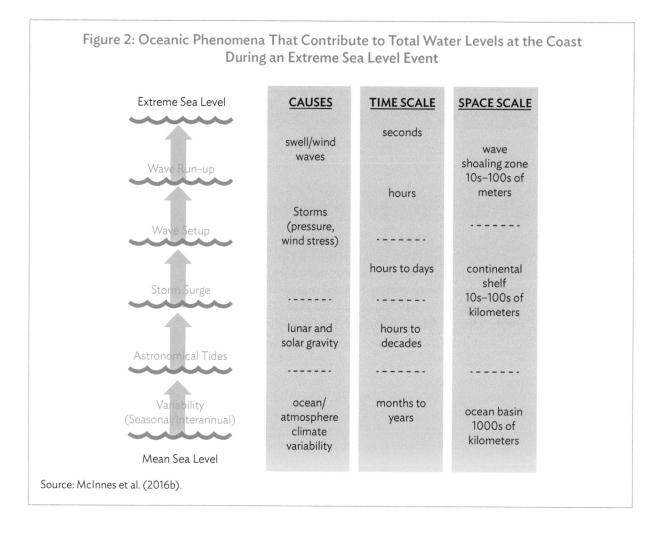

Figure 2: Oceanic Phenomena That Contribute to Total Water Levels at the Coast During an Extreme Sea Level Event

Source: McInnes et al. (2016b).

B.   Weather and Climate Influences on Sea-Level Change in the Pacific Islands Region

Numerous weather and climate processes drive sea-level variability and change in the PIR. Complex interactions between the ocean and atmosphere result in the continual exchange of heat and water, driven by prevailing winds to achieve dynamic and/or thermodynamic equilibrium. Figure 3 summarizes the weather and climate processes that influence the PIR from November to April (the tropical cyclone season when storm surge and other short-term SLR events are most common). These climate influences operate over different temporal and spatial scales, from intra-seasonal to interdecadal. Given the spatial extent of the PIR, different climate phenomena can influence locations within the PIR in various ways.

Tropical cyclones account for 76% of disasters in the PIR and bring extreme winds, waves, intense storm surge, and prolonged rainfall with fluvial, pluvial, and coastal flooding (Terry et al. 2004; McInnes et al. 2011; Brown et al. 2016). For example, Tropical Cyclone Tomas (March 2010) generated water levels associated with wave run-up and storm surge of 7 m above mean sea level around the Lau Island Group in Fiji (Needham et al. 2015). Tropical cyclone–induced waves and storm surges can be observed at vast distances from the system itself (e.g., Tropical Cyclone Pam generated large waves and significant storm surge that impacted Tuvalu, about 1,100 kilometers [km] to the northeast of the storm system) (Hoeke et al. 2021). Tropical cyclones produce significant wave heights, which lead to extreme coastal water levels as a result of wave setup and run-up for unsheltered locations. For instance, Tropical Cyclone Ofa (February 1990) generated a storm surge of 1.6 m with significant wave heights of 8.1 m south of Samoa and up to 18 m on the island of Niue (Solomon and Forbes 1999).

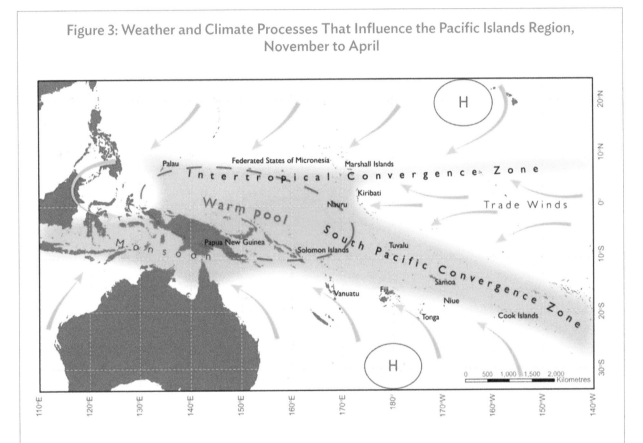

Figure 3: Weather and Climate Processes That Influence the Pacific Islands Region, November to April

Note: Yellow arrows indicate near-surface winds. The red-dashed region indicates the typical location of the Pacific Warm Pool (i.e., during years classed as the neutral phase of the El Niño–Southern Oscillation). High-pressure systems are indicated by H.

Source: https://www.pacificclimatechangescience.org/wp-content/uploads/2013/06/Climate-in-the-Pacific-summary -48pp_WEB.pdf. Australian Bureau of Meteorology and Commonwealth Scientific and Industrial Research Organisation (2014).

Coupled ocean-atmosphere modes such as ENSO cause substantial seasonal to interannual variations in sea level across the PIR (e.g., Church et al. 2006). During La Niña events, strengthened westerly winds and intensification of the Walker circulation displace the Pacific Warm Pool toward the west, resulting in sea levels that are 20–30 cm higher than normal in the western PIR (e.g., Becker et al. [2012]). Since 1970, evidence has pointed to potential intensification of La Niña–related sea-level anomalies (Becker et al. 2012). Conversely, El Niño results in the displacement of the Pacific Warm Pool toward the east, resulting in sea levels that are 20–30 cm lower than normal in the western PIR. During El Niño events, increased wave heights and an anticlockwise rotation of wave direction are typically observed in the eastern equatorial Pacific (Hemer et al. 2011).

Other aspects of ENSO influence sea levels in the PIR. ENSO is typically characterized by anomalously warm sea surface temperatures (SSTs) (for El Niño) or cool SSTs (for La Niña) in the eastern equatorial Pacific. El Niño

Modoki (also known as the Central Pacific El Niño or the Dateline El Niño) (Ashok et al. 2007) is associated with anomalous SST warming and associated SLR in the central tropical Pacific and anomalous SST cooling in the eastern and western tropical Pacific. Conversely, La Niña Modoki is associated with anomalous SST cooling in the central tropical Pacific and anomalous SST warming and associated SLR in the eastern and western tropical Pacific. El Niño Modoki conditions have become more frequent since 2002, resulting in increased sea levels in the central PIR (Becker et al. 2012).

Interdecadal climate variability also influences sea levels across the PIR, particularly via the modulation of the frequency of El Niño and La Niña impacts during different IPO phases (Kiem et al. 2003; Magee et al. 2017). Since negative IPO phases are associated with more La Niña events than positive IPO phases, negative IPO epochs (e.g., ~1945–1976, ~1999–present) are associated with an intensification and expansion of the elevated sea levels in the western PIR that are typical during La Niña events (Becker et al. 2012). This observation is supported by reports of increases in non-tropical cyclone–related coastal inundation events in the western Pacific (e.g., Kiribati, Tuvalu, and the Marshall Islands) since 1998 (Ramsay 2011).

C.   Terrestrial Factors That Influence Relative Sea-Level Change in the Pacific Islands Region

The impacts of SLR (e.g., coastal flooding, inundation, erosion, soil and water salinization) are also influenced by terrestrial factors such as vertical land movement, geomorphology, and sediment availability (McInnes et al. 2016b; Masselink et al. 2020). In addition to long-term vertical land movement from glacial isostatic adjustment, local vertical land movement caused by tectonic activity and volcanism has been found to produce larger vertical movements than decadal changes in absolute sea level, particularly in Tonga and Vanuatu, which are close to active plate boundaries (Ballu et al. 2011). For example, in April 1997, a magnitude 7.8 earthquake near the Torres Islands, Vanuatu, caused abrupt subsidence of 0.5–1 m, contributing to a significant rise in relative sea level and an increase in coastal inundation and flooding extent over subsequent years (Ballu et al. 2011; Ramsay 2011).

Vertical land movement also occurs because of human activities such as gas and groundwater extraction, urbanization, and sediment consolidation (Becker et al. 2012; McInnes et al. 2016b) or vertical island accretion (Masselink et al. 2020). Vertical movements in the PIR are monitored using tide-gauge records and Global Positioning System (GPS) technologies. However, fewer than 20 inhabited islands in the PIR have such systems in place to monitor land movement.

Analysis of land-based global navigation satellite system (GNSS) stations provides insights into vertical land movement in the PIR (Table 1). These records are typically short (from 4.5 years for French Polynesia [Tubuai] to 17 years for New Caledonia [Lifou]) and many are not continuous. Most stations listed in Table 1 are subsiding, amplifying the impact of increases in relative sea levels, particularly in the Torres Islands (Vanuatu), where absolute sea level rose by 150 ± 20 millimeters (mm) in 1997–2009. However, GPS data reveal that some sites in the Torres Islands subsided by up to 117 ± 30 mm over the same period, almost doubling the effective SLR (see Ballu et al. [2011] for further information).

Figure 4 summarizes work by Geoscience Australia (Brown et al. 2020), which comprehensively assessed the absolute vertical rate of movement (the rate at which land moves up or down with respect to the center of the Earth) over 2003–2018 using 13 Pacific Island tide gauges from the Pacific Sea Level and Geodetic Monitoring Project. There are ongoing efforts by the Climate and Ocean Services Program in the Pacific to integrate vertical land movement data with sea-level tide data in the PIR since the early 1990s. Like locations listed in Table 1, most of those assessed by Brown et al. (2020) are subsiding, thereby amplifying the impacts of SLR.

Table 1: Vertical Land Movements in the Pacific Islands Region as Measured by Global Navigation Satellite System Stations

Station	Period of Data (month/year)	Time Span (year)	Data Availability (% complete)	Total Vertical Land Movement (millimeter)	Vertical Land Movement (millimeter/year)
Cook Islands (Rarotonga)	9/2001–6/2019	12.30	75.02	–6.15 ± 4.43	–0.50 ± 0.36
Fiji (Lautoka)	11/2001–9/2019	12.10	85.00	–13.92 ± 3.15	–1.15 ± 0.26
French Polynesia (Tubuai)	5/2009–10/2017	4.53	84.52	–1.49 ± 2.36	–0.33 ± 0.52
French Polynesia (Papeete)	12/2003–9/2019	9.96	83.24	–19.32 ± 2.29	–1.94 ± 0.23
French Polynesia (Tahiti-Faaa)	10/2006–9/2019	7.20	71.05	–12.96 ± 3.24	–1.80 ± 0.45
French Polynesia (Rikitea)	4/2000–9/2019	10.49	45.97	–10.8 ± 4.20	–1.03 ± 0.40
Futuna	9/1998–9/2019	15.26	57.78	–4.27 ± 3.97	–0.28 ± 0.26
Kiribati (Betio Island)	7/2002–9/2019	11.41	85.72	–2.51 ± 2.74	–0.22 ± 0.24
Nauru	6/2003–9/2019	10.50	76.18	–10.08 ± 2.63	–0.96 ± 0.25
New Caledonia (Noumea)	7/1997–3/2007	9.28	96.31	–12.99 ± 2.60	–1.40 ± 0.28
New Caledonia (Noumea)	5/2006–9/2019	7.62	92.02	–14.17 ± 1.75	–1.86 ± 0.23
New Caledonia (Lifou)	3/1996–9/2019	17.40	91.12	2.96 ± 7.66	0.17 ± 0.44
New Caledonia (Koumac)	4/1996–9/2019	17.69	88.86	–2.83 ± 3.18	–0.16 ± 0.18
Papua New Guinea (Lae)	1/2001–8/2019	11.30	56.13	–57.44 ± 3.06	–5.07 ± 0.27
Papua New Guinea (Manus Island)	5/2002–9/2019	11.66	80.83	–31.37 ± 5.25	–2.69 ± 0.45
Samoa				Unreliable because of station problems	
Solomon Islands (Honiara)				Variable and/or unreliable because earthquakes are common	
Tonga (Nukualofa)	2/2002–9/2019	11.86	80.90	35.7 ± 4.86	3.01 ± 0.41
Tuvalu (Funafuti)	11/2001–2/2019	12.05	70.28	–20.61 ± 2.05	–1.71 ± 0.17
Vanuatu (Torres Islands)	1/1997–12/2009	13.00	94.00	–117 ± 30	–9.00 ± 3.33

Source: Sonel. Vertical Land Movements. https://www.sonel.org/-Vertical-land-movement-estimate-.html?lang=en.

Table 2 compares the results from Figure 4 (Brown et al. 2020) with the vertical land movement estimates used in AR6 (Fox-Kemper et al. 2021). Again, the message is that subsidence is occurring in most PIR locations, and this is amplifying the impacts of SLR.

Figure 4: Absolute Vertical Rate of Movement of Tide Gauges in Pacific Island Countries, 2003–2018

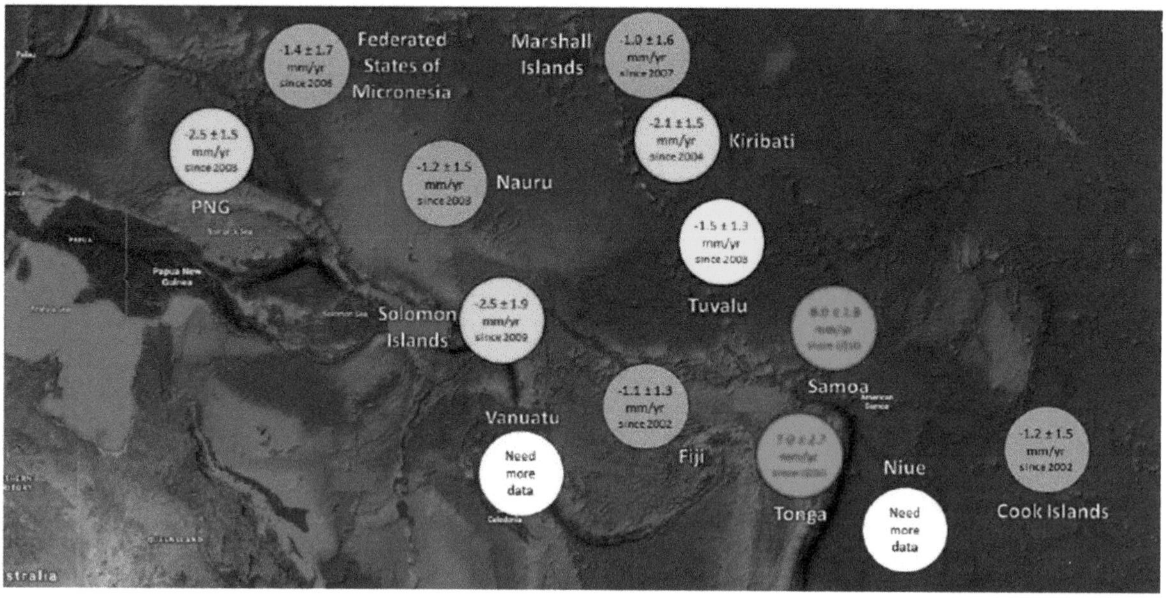

PNG = Papua New Guinea.

Notes: Negative values denote subsiding land. Yellow circles represent sites with 0–2 mm/year of land subsidence. Orange circles represent sites with 2–4 mm/year of land subsidence. Red circles represent sites with greater than 4 mm/year of land subsidence. Gray circles represent sites with an absolute vertical rate of movement within the range of data uncertainty. In these cases, either the absolute vertical rate of movement of the tide gauge is close to zero, or a longer time series of data is needed to better understand the absolute vertical rate of movement of the tide gauge.

Source: Brown et al. (2020), Figure 8.

Table 2: Absolute Vertical Rate of Movement of the Tide Gauges in Pacific Island Countries

Location	AR6 Vertical Movement (mm/year)[a]	Gauge Vertical Movement (mm/year)[b]
Cook Islands	0.1 (-1.0/1.2)	-1.2 ±1.5 (2002)
FSM	-0.5 (-1.3/0.2)	-1.4 ±1.7 (2006)
Samoa	-1.4 (-2.1/-0.7)	-8.0 ±2.8 (2010)
Tonga	-0.3 (-1.4/0.8	-7.0 ±2.7 (2010)
Niue	-0.2 (-1.8/1.5)	~-1.5 (2006)
Fiji	0.0 (1.0/0.9)	-1.1 ±1.3 (2002)
Kiribati	-0.1 (1.3/0.9)	-2.1 ±1.5 (2004)
Nauru	-0.1 (-1.2/1.0)	-1.2 ±1.5 (2003)
Palau	0.0 (-1.0/1.0)	–
Marshall Islands	-0.6 (-1.3/0.1)	-1.0 ±1.6 (2007)
Vanuatu	0.3 (-0.8/1.4)	~3.0 (2013)
Solomon Islands	0.5 (-0.3/1.4)	-2.5 ±1.9 (2009)
Tuvalu	-0.3 (-1.1/0.5)	-1.5 ±1.3 (2003)

AR6 = Sixth Assessment Report of the Intergovernmental Panel on Climate Change, FSM = Federated States of Micronesia, GNSS = global navigation satellite system.

Notes: Results from Fox-Kemper et al. (2021) (Sixth Assessment Report of the Intergovernmental Panel on Climate Change) compared with results from Brown et al. (2020), Figure 4. Negative values imply subsidence and positive values uplift.

[a] Absolute vertical rate of land movement based on AR6 long-term trends as presented in Fox-Kemper et al. (2021). Rates defined as 50%ile and *most likely* range (5%/95%) shown in brackets.

[b] Local vertical land movement at the Pacific tide gauges with start of dataset shown in brackets. Rates are averages based on GNSS measurements at the tide gauge from Brown et al. (2020). Estimates (~) are based on movement of GNSS base station.

Source: Pacific Region Infrastructure Facility (2021), Table 5.5.

III. Historical Sea-Level Change in the Pacific Islands Region

A. Instrumental Period (since ~1900)

AR6 reports median global mean sea-level change of +1.7 mm/year (range is 1.3–2.2 mm/year) from 1901–2018, +2.3 (1.6–3.1) mm/year from 1971–2018, and +3.7 (3.2–4.2) mm/year from 2006–2018 (Intergovernmental Panel on Climate Change [IPCC] 2021). This acceleration in global SLR has been attributed to increased rates of ocean warming (and associated expansion) and terrestrial ice melt (Bindoff et al. 2007; Nurse et al. 2014; IPCC 2019, 2021). However, SLR is not spatially uniform, and sea level in much of the PIR is reported to have risen at a rate three to four times greater than the global mean (Cazenave and Llovel 2009; Nerem et al. 2010; Wang et al. 2021) (Figure 5). In Nauru, Funafuti, Pago Pago, Papeete, Noumea, and Tarawa, the SLR trend is significantly greater than the global average SLR trend. For example, in Funafuti, sea-level trends over the last 60 years have been ~5 mm/year, leading to a total sea rise of ~30 cm since 1950 (Becker et al. 2012). These particularly high rates of SLR are likely a combination of an underlying rising trend that is magnified in some locations by the impacts of natural climate variability (e.g., La Niña–dominated IPO negative phase since ~1999 [see page 6, second paragraph] and other local and regional influences [as summarized in sections II-B and II-C]).

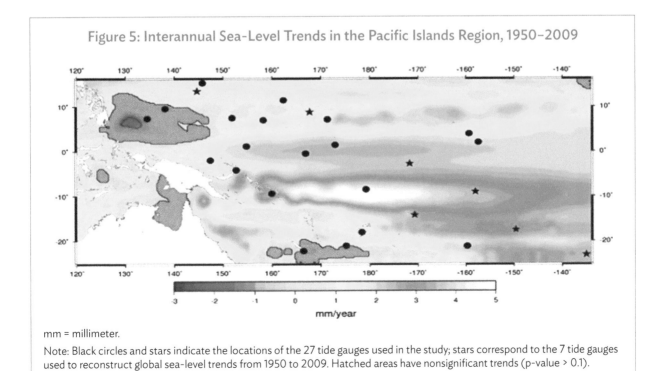

Figure 5: Interannual Sea-Level Trends in the Pacific Islands Region, 1950–2009

mm = millimeter.

Note: Black circles and stars indicate the locations of the 27 tide gauges used in the study; stars correspond to the 7 tide gauges used to reconstruct global sea-level trends from 1950 to 2009. Hatched areas have nonsignificant trends (p-value > 0.1).

Source: Becker et al. (2012).

B. Pre-Instrumental Period (before ~1950)

Although there is a tendency to focus on future sea-level change, it is important to recognize that sea levels (and rates of change) higher than those observed in the instrumental period (~1950 to the present) occurred in the pre-instrumental period. For example, ~12,000–15,000 years ago, SLR of 5 m per century is evident in some paleoclimate records (e.g., Fairbanks 1989; Deschamps et al. 2012). Hansen et al. (2016) claim that during the last interglacial period (~130,000–116,000 years before 1950), when temperatures were less than 1°C warmer than now, sea levels were 6–9 m higher than they were in 2020. Although insights from paleoclimate records are useful for demonstrating that ice sheets can respond rapidly and can produce dramatic rates of SLR, no past period is a perfect analogue for the current or future climate situation. The Hansen et al. (2016) assessment is associated with major caveats and has certainly not been accepted unquestionably. For example, the pace at which such extreme SLR (i.e., several meters) might occur depends on the relative configuration of ice sheets and the extent to which they have reached critical tipping points (among other factors) (Thorne 2015). Hence, the current consensus in the literature is that such extreme SLR would happen over long periods (centuries to millennia) and is unlikely to occur before 2100 (IPCC 2019, 2021). Chapter 9 in AR6 states that projected global mean SLR of 1.7–6.8 m by 2300 is possible without marine ice cliff instability and this projection increases to 16 m by 2300 with marine ice cliff instability (Fox-Kemper et al. 2021).

IV. Sea-Level Change Projections for the Pacific Islands Region

A. Mean Sea-Level Change Projections for the Pacific Islands Region

AR6 (IPCC 2021) used output from the latest generation of global climate models, produced as part of the sixth Coupled Model Intercomparison Project. These coordinated efforts consist of simulations of ~100 distinct climate models developed by different research groups. AR6 used a new set of scenarios derived from the Shared Socioeconomic Pathways (SSPs) (Figure 6). The SSPs consist of five broad narratives of future socioeconomic development that define scenarios of energy use, air pollution control, land use, and greenhouse gas emissions to which Representative Concentration Pathways (RCPs) are applied to achieve approximate radiative forcing levels to the end of the 21st century. This is in contrast to AR5, which was based on a fixed socioeconomic pathway whereby various RCPs were applied. The AR6 SSP suite is considered more representative of potential climate futures. AR6 also introduced a new mean sea-level baseline period of 1995–2014 to which the projections are referenced (AR5 used 1986–2005 as the baseline period).

AR6 projections (IPCC 2021) include research that emerged after AR5 on the role and contribution of ice sheet melt. This research suggests that global mean sea level will continue to rise throughout the 21st century and beyond (Figures 7–9 and Table 3). The AR6 assessment is broadly consistent with the IPCC Special Report on the Oceans and Cryosphere in a Changing Climate (SROCC) (IPCC 2019) and AR5 (Church et al. 2013) assessments (Figure 9), but AR6 projections contain nearly twice the amount of SLR because of Antarctic melting than previous assessments (Figure 7), resulting in slightly higher projections of SLR to 2100 (Figure 9).

Figure 6: Shared Socioeconomic Pathways Used in the Sixth Assessment Report of the Intergovernmental Panel on Climate Change

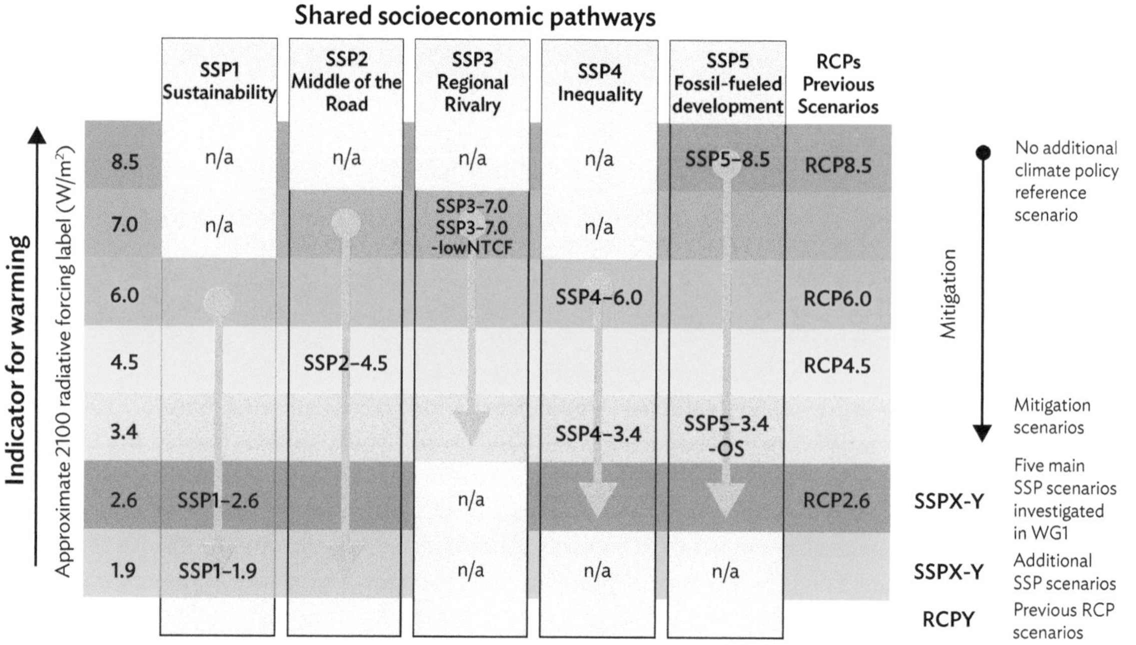

AR5 = Fifth Assessment Report of the IPCC, IPCC = Intergovernmental Panel on Climate Change, NTCF = Near-Term Climate Forcer, OS = Overshoot Scenario, RCP = Representative Concentration Pathway, SSP = Shared Socioeconomic Pathway, WG1 = Working Group 1, W/m^2 = watt per square meter.

Notes:
1. The figure shows the SSPs used in the Sixth Assessment Report of the IPCC, with the radiative forcing categorization and SSP storylines upon which they were built.
2. **Very low emissions scenario** (SSP1–1.9). Holds warming to ~1.5°C above 1850–1900 temperatures in 2100 "after slight overshoot" and implied net-zero CO_2 emissions around the middle of the century.
3. **Low emissions scenario** (SSP1–2.6). Stays below 2°C warming, with implied net-zero emissions in the second half of the century. Most consistent with AR5 RCP2.6.
4. **Intermediate emissions scenario** (SSP2–4.5). Approximately in line with the upper end of combined pledges under the Paris Agreement. The scenario deviates mildly from a "no additional climate policy" reference scenario, resulting in a best-estimate warming of ~2.7°C by 2100. Most consistent with AR5 RCP4.5.
5. **High emissions scenario** (SSP3–7.0): A medium- to high-reference scenario resulting from no additional climate policy, with particularly high non–CO_2 emissions, including high aerosol emissions.
6. **Very high emissions scenario** (SSP5–8.5). A high-reference scenario with no additional climate policy. Emissions as high as SSP5–8.5 are only achieved within the fossil-fueled SSP5 socioeconomic development pathway. Most consistent with AR5 RCP8.5.

Source: IPCC (2021), Cross-Chapter Box 1.4, Figure 1.

Figure 7: AR6 Median Global Mean and Regional Relative Sea-Level Change Projections (meters) by Contribution Under the SSP1-5 2.6 and SSP5-8.5 Scenarios

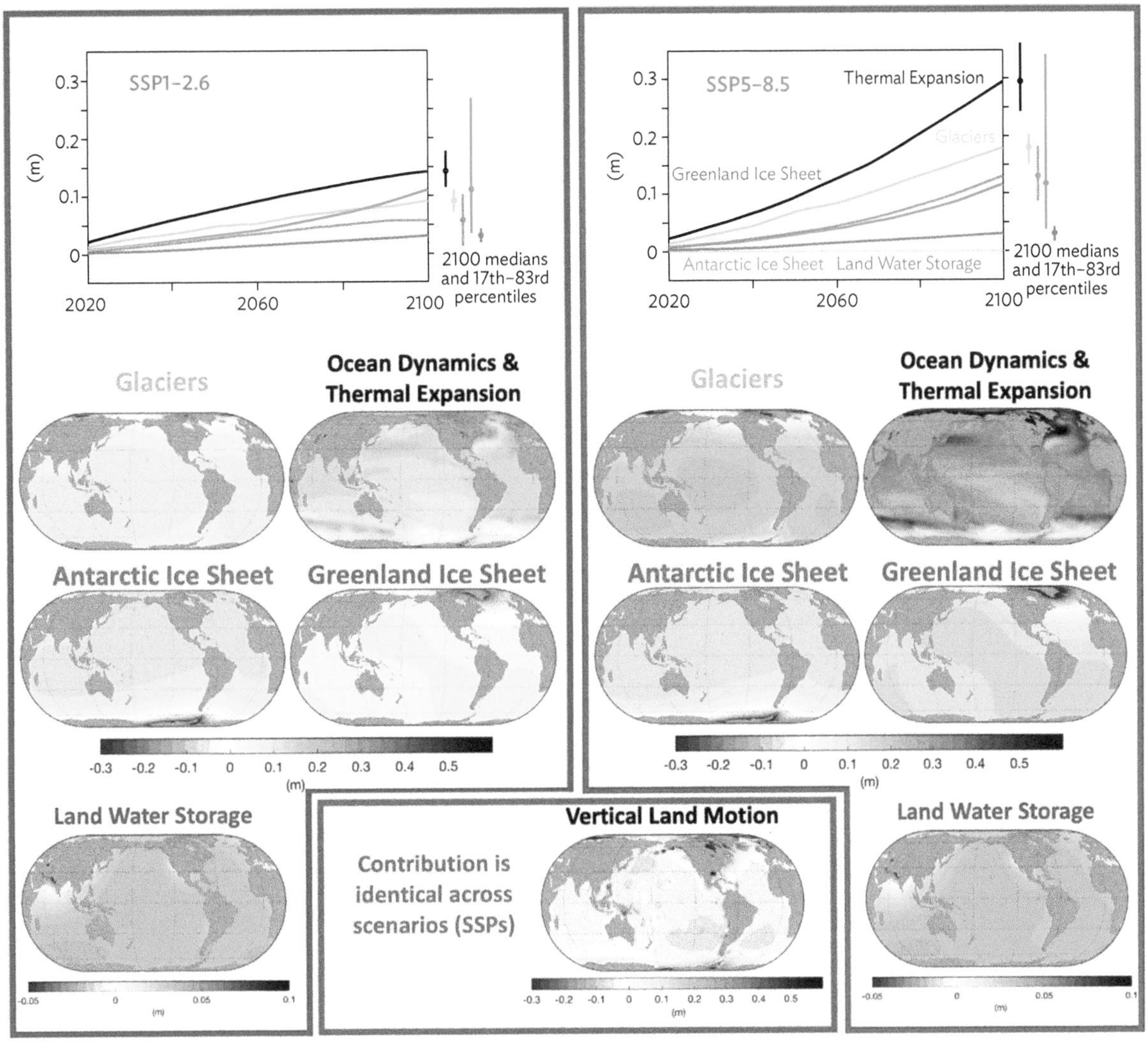

AR6 = Sixth Assessment Report of the IPCC, IPCC = Intergovernmental Panel on Climate Change, SSP = Shared Socioeconomic Pathway.

Notes:
1. Upper time series: Global mean contributions to sea-level change as a function of time, relative to 1995–2014.
2. Lower maps: Regional projections of the sea-level contributions in 2100 relative to 1995–2014 for SSP5-8.5 and SSP1-2.6.
3. Vertical land motion is common to both SSPs.

Source: IPCC (2021), Chapter 9, Figure 9.26.

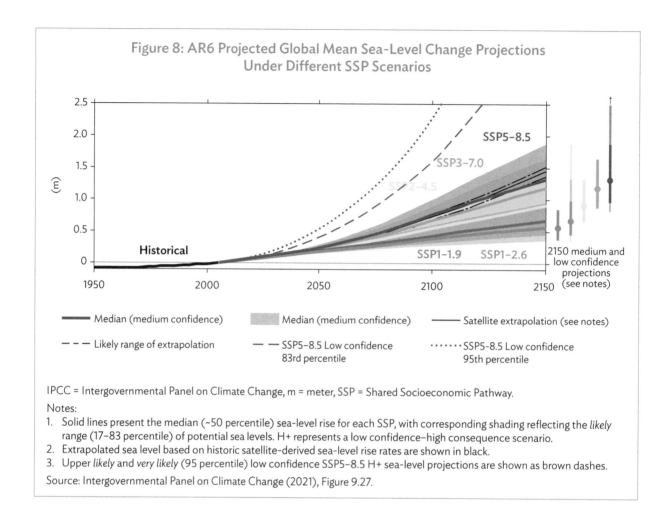

Figure 8: AR6 Projected Global Mean Sea-Level Change Projections Under Different SSP Scenarios

IPCC = Intergovernmental Panel on Climate Change, m = meter, SSP = Shared Socioeconomic Pathway.

Notes:
1. Solid lines present the median (~50 percentile) sea-level rise for each SSP, with corresponding shading reflecting the *likely* range (17–83 percentile) of potential sea levels. H+ represents a low confidence–high consequence scenario.
2. Extrapolated sea level based on historic satellite-derived sea-level rise rates are shown in black.
3. Upper *likely* and *very likely* (95 percentile) low confidence SSP5–8.5 H+ sea-level projections are shown as brown dashes.

Source: Intergovernmental Panel on Climate Change (2021), Figure 9.27.

Table 3: Global Median Sea-Level Change Projections (meters) to 2150 (relative to the 1995–2014 mean sea-level baseline) for Selected Shared Socioeconomic Pathways (SSPs)

Year	Low Emissions (SSP1–2.6)	Intermediate Emissions (SSP2–4.5)	High Emissions (SSP3–7.0)	Very High Emissions (SSP5–8.5)	Very High Emissions–Low (SSP5–8.5 H+)
2030	0.09 (0.08–0.12)	0.09 (0.08–0.12)	0.09 (0.08–0.12)	0.10 (0.09–0.12)	0.10 (0.09–0.15)
2050	0.19 (0.16–0.25)	0.20 (0.17–0.26)	0.21 (0.18–0.27)	0.23 (0.20–0.29)	0.24 (0.20–0.40)
2090	0.39 (0.30–0.54)	0.48 (0.38–0.65)	0.56 (0.46–0.74)	0.63 (0.52–0.83)	0.71 (0.52–1.30)
2100	0.44 (0.32–0.61)	0.56 (0.43–0.76)	0.68 (0.55–0.90)	0.77 (0.63–1.01)	0.88 (0.63–1.60)
2150	0.68 (0.46–0.99)	0.92 (0.66–1.33)	1.19 (0.89–1.65)	1.32 (0.98–1.88)	1.98 (0.98–4.82)

SSP = Shared Socioeconomic Pathway.

Notes:
1. Very High Emissions–Low (SSP5–8.5 H+) represents a low confidence–high consequence scenario.
2. Bracketed values show *likely* range (17–83 percentile).

Source: Intergovernmental Panel on Climate Change (2021), Table 9.9.

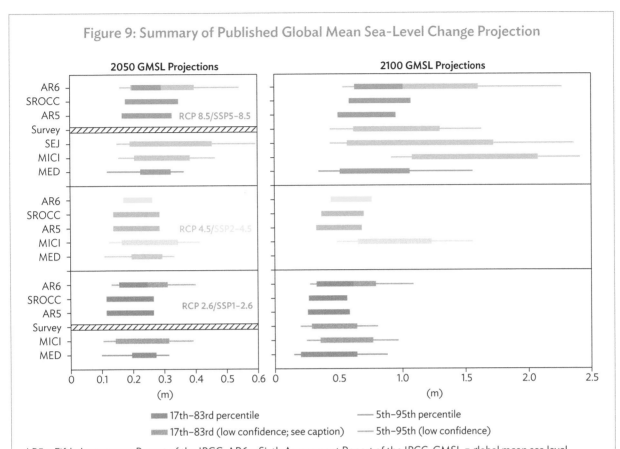

Figure 9: Summary of Published Global Mean Sea-Level Change Projection

AR5 = Fifth Assessment Report of the IPCC, AR6 = Sixth Assessment Report of the IPCC, GMSL = global mean sea level, IPCC = Intergovernmental Panel on Climate Change, m = meter, MED = medium confidence, MICI = marine ice cliff instability, RCP = Representative Concentration Pathway, SEJ = structured expert judgement, SROCC = Special Report on the Ocean and Cryosphere in a Changing Climate (IPCC 2019), SSP = Shared Socioeconomic Pathway.

Notes: Projections are relative to the 1995–2014 baseline and standardized to account for differences in time periods. Thick bars span the 17th to the 83rd percentile projections, and thin bars span the 5th to the 95th. The different assessments of ice sheet contributions are indicated by "MED" (ice sheet projections including only processes in whose quantification there is medium confidence), "MICI" (ice sheet projections which incorporate marine ice cliff instability), and "SEJ" (structured expert judgement to assess the central range of the ice-sheet projection distributions). "Survey" indicates the results of a 2020 survey of sea-level experts on GMSL rise from all sources (Horton et al. 2020). Projection categories incorporating processes in which there is low confidence (MICI and SEJ) are lightly shaded. Dispersion among the different projections represents deep uncertainty, which arises as a result of low agreement on the appropriate conceptual models describing ice sheet behavior and low agreement on probability distributions used to represent key uncertainties.

Source: IPCC (2021), Figure 9.25.

As per historical trends (section III), Figure 10 shows that AR6 SLR projections for the PIR are at (or above) the upper bound of the projections for global mean SLR and that a decrease in sea level is highly unlikely for the PIR. Regardless of the SSP scenario, higher amounts of SLR are expected in the North and South Pacific than in the equatorial region because of the projected changes to ENSO.

As part of the Pacific-Australia Climate Change Science and Adaptation Planning (PACCSAP, https://www.pacificclimatechangescience.org/) program, regionally specific SLR projections were calculated from AR5 climate modeling for 14 island nations and territories across the PIR (Australian Bureau of Meteorology and Commonwealth Scientific and Industrial Research Organisation, 2014). The PACCSAP

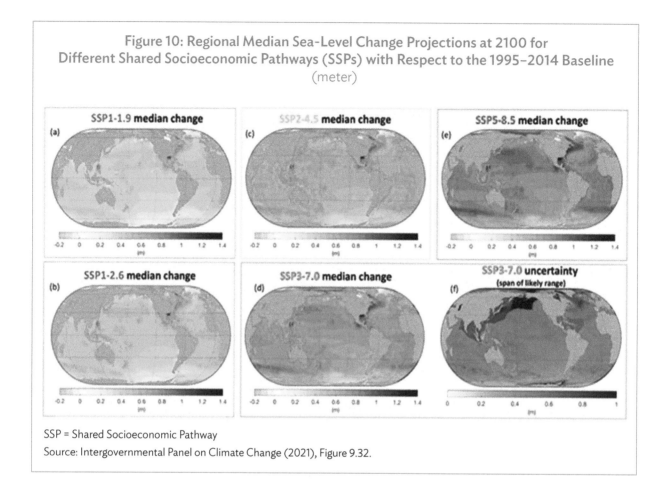

Figure 10: Regional Median Sea-Level Change Projections at 2100 for Different Shared Socioeconomic Pathways (SSPs) with Respect to the 1995–2014 Baseline (meter)

SSP = Shared Socioeconomic Pathway

Source: Intergovernmental Panel on Climate Change (2021), Figure 9.32.

program information is summarized in Figure 11 and Figure 12 and represents the most up-to-date, location-specific analysis of SLR projections for the PIR. Further information from the PACCSAP program on location-specific changes in SLR in the PIR for four AR5 RCPs (RCP2.6, RCP4.5, RCP6.0, RCP8.5) and for four periods (20 years centered on 2030, 2050, 2070, and 2090) is included in Appendix 1. Work is underway to update the PACCSAP program location-specific SLR projections for the PIR to incorporate the AR6 science and SLR modeling. In the meantime, localized versions of the AR6 projections can be viewed at the United States National Aeronautics and Space Administration (NASA) and IPCC Sea Level Projection Tool.

Figure 11, Figure 12, and Appendix 1 demonstrate that although there is very high confidence in the direction of change for all PIR locations (i.e., decrease in sea level is not projected anywhere in the PIR), there is only medium confidence in the magnitude of change. This is due to uncertainties associated with (i) projections of Antarctic ice sheet contributions (Slater et al. 2021); (ii) the influence of natural interannual to decadal variability, which could lead to conditions where sea levels are further elevated (e.g., because of increased tropical cyclones [see page 4, second paragraph] or increased by La Niña [see page 6, second paragraph]; see Table 4 for an indication of how natural climate variability could amplify SLR in the PIR); and (iii) the gravitational fingerprint associated with global redistribution of water from Greenland and Antarctic ice melt (Bamber and Riva 2010; Kopp et al. 2014; Hsu and Velicogna 2017). Despite these (and other) uncertainties, a comparison of the information in Appendix 1 with the global IPCC projections (Figures 7–9 and Table 3) shows (i) strong regional variations in SLR across ocean basins; and (ii) that for each RCP, the SLR projections for all islands in the PIR are at or above the upper range of the IPCC global mean SLR projections (both these points are also confirmed by Wang et al. (2021)).

It is also important to note that the SLR projections shown in Figure 11, Figure 12, and Appendix 1 do not consider the influence of changes in storm surge and/or wave climate (power and direction). The potential influence of changes to storm surge, wave climate, and/or wave power are discussed in section IV-C, section IV-D, and section IV-E, respectively.

The SLR projections shown in Figure 11, Figure 12, and Appendix 1 also do not consider the impact of vertical land movement, which (as per section II-C) can amplify or reduce the local impacts of SLR.

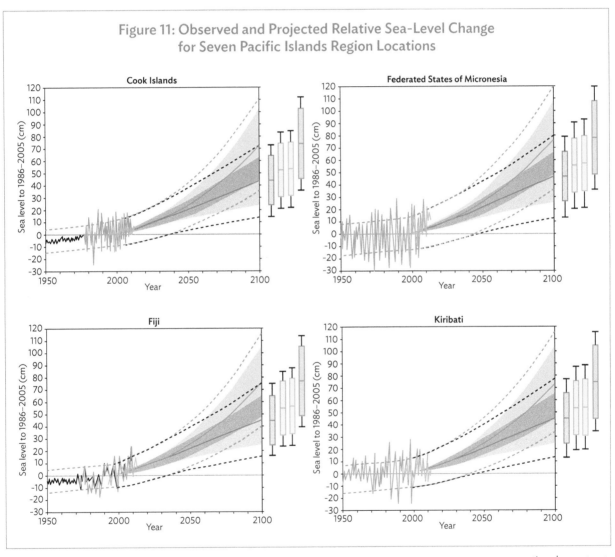

Figure 11: Observed and Projected Relative Sea-Level Change for Seven Pacific Islands Region Locations

continued on next page

Figure 11 continued

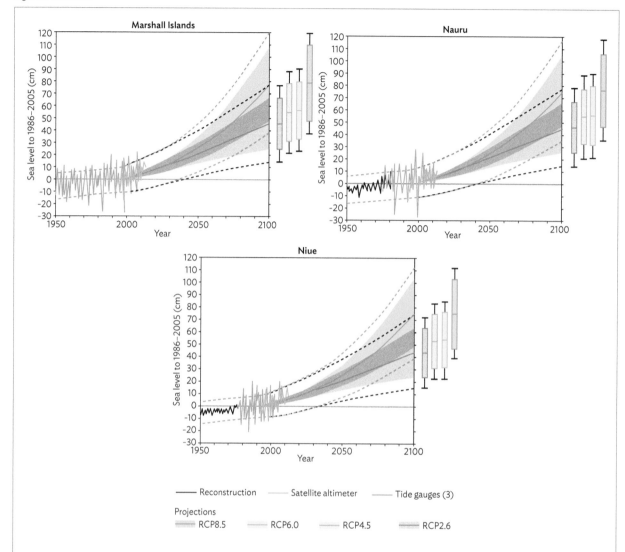

cm = centimeter, RCP = Representative Concentration Pathway.

Notes: The observed tide gauge records of relative sea level since the late 1970s are indicated in purple, and the satellite record since 1993 in green. Reconstructed sea level since 1950 is shown in black. Multimodal mean projections in 1995–2100 are given for RCP8.5 (red solid line) and RCP2.6 (blue solid line), with the 5%–95% uncertainty range shown by the red- and blue-shaded regions. The ranges of projections for four emission scenarios (RCP2.6, RCP4.5, RCP6.0, RCP8.5) by 2100 are shown by the bars on the right. The dashed lines are an estimate of interannual variability in sea level (5%–95% uncertainty range about the projections) and indicate that individual monthly averages of sea level can be above or below longer-term averages.

Source: Australian Bureau of Meteorology and Commonwealth Scientific and Industrial Research Organisation (2014).

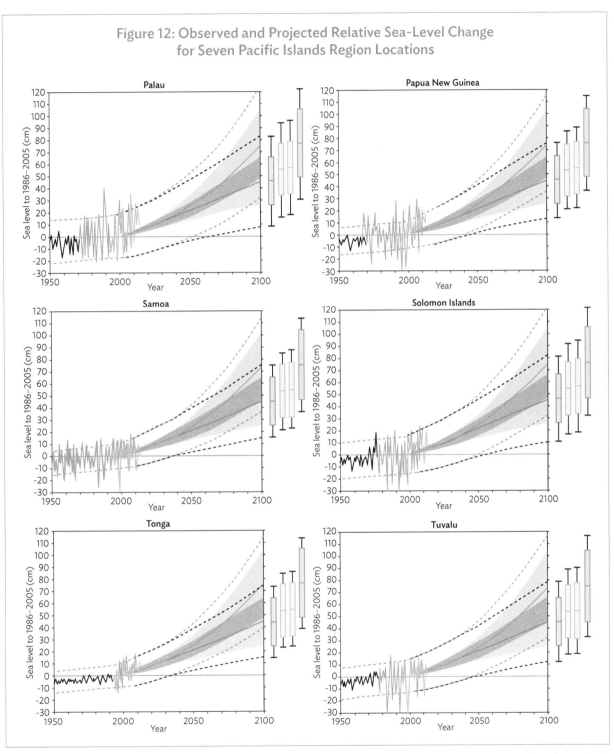

Figure 12: Observed and Projected Relative Sea-Level Change
for Seven Pacific Islands Region Locations

continued on next page

Figure 12 continued

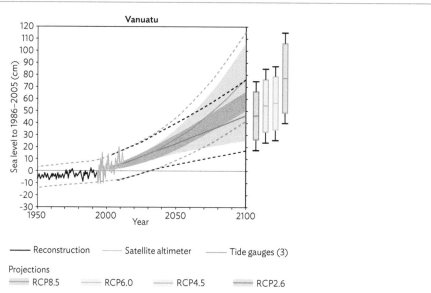

cm = centimeter, RCP = Representative Concentration Pathway.

Note: The observed tide gauge records of relative sea level since the late 1970s are indicated in purple, and the satellite record since 1993 in green. Reconstructed sea level since 1950 is shown in black. Multimodal mean projections in 1995–2100 are given for RCP8.5 (red solid line) and RCP2.6 (blue solid line), with the 5%–95% uncertainty range shown by the red- and blue-shaded regions. The ranges of projections for four emission scenarios (RCP2.6, RCP4.5, RCP6.0, RCP8.5) by 2100 are shown by the bars on the right. The dashed lines are an estimate of interannual variability in sea level (5%–95% uncertainty range about the projections) and indicate that individual monthly averages of sea level can be above or below longer-term averages.

Source: Australian Bureau of Meteorology and Commonwealth Scientific and Industrial Research Organisation (2014).

Table 4: Projected Changes in Sea-Level Rise under RCP8.5 by 2090 for 14 Locations in the Pacific Islands Region
(centimeter)

	Projected SLR (from Appendix 1)[a]	Historical Interannual Variability	Upper Bound of Projected SLR plus Historical Interannual Variability
Cook Islands	39–86	19	105
Federated States of Micronesia	41–90	26	116
Fiji	41–88	18	106
Kiribati	38–87	23	110
Marshall Islands	41–92	20	112
Nauru	41–89	23	112
Niue	41–87	17	104
Palau	41–88	36	124
Papua New Guinea	47–87	23	110
Samoa	40–87	20	107
Solomon Islands	40–89	31	120
Tonga	41–88	18	106
Tuvalu	39–87	26	113
Vanuatu	42–89	18	107

RCP = Representative Concentration Pathway, SLR = sea-level rise.

[a] Values represent the 5%–95% range (see Appendix 1 for further details).

Note: Upper bound of projected SLR includes the influence of historical interannual variability. Historical interannual variability is taken from dashed lines in Figure 11 and Figure 12 (5%–95% range, after removal of the seasonal signal).

Source: Australian Bureau of Meteorology and Commonwealth Scientific and Industrial Research Organisation (2014).

B.   Global Sea-Level Projections and Their Relevance to the Pacific Islands Region

The previous section summarized evidence about the projected impacts of anthropogenic climate change on sea levels. A key finding was that there is very high confidence in the direction of change but large uncertainty associated with the projected magnitude and timing of SLR. As discussed in Kopp et al. (2017), the upper bounds of future SLR projections remain deeply uncertain. Moreover, Garner et al. (2018: page 1611) suggested that "the deeply uncertain nature of global SLR projections is evident by the fact that there is no unique probability distribution of future sea level; thus, it is unlikely that there will be any particular method that is found to be best for estimating future sea level change anytime in the near future." Therefore, Garner et al. (2018) compared more than 70 global SLR studies conducted during 1983–2018 to gain insights into the range of 21st century global SLR considered plausible, and how that range has changed as science has evolved (Figure 13).

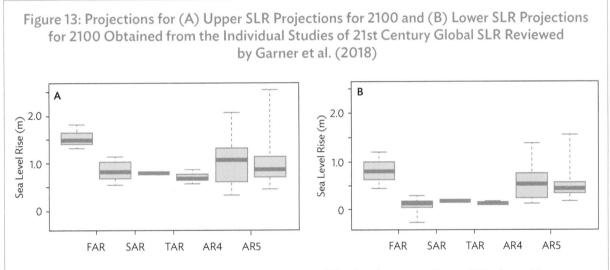

Figure 13: Projections for (A) Upper SLR Projections for 2100 and (B) Lower SLR Projections for 2100 Obtained from the Individual Studies of 21st Century Global SLR Reviewed by Garner et al. (2018)

AR4 = Fourth Assessment Report, AR5 = Fifth Assessment Report, FAR = First Assessment Report, SAR = Second Assessment Report, TAR = Third Assessment Report.

Notes: Box edges are the interquartile (25th to 75th percentiles) range; solid lines are the 50th percentile. Whiskers extend to data extremes (0 to 100th percentiles) to show the full range of SLR projections in each case. The horizontal axis uses the Intergovernmental Panel on Climate Change (IPCC) assessment reports to divide the literature based on publication date.

Source: Garner et al. 2010.

Figure 13 shows that later research has tended to raise the upper uncertainty bound as more is learned about specific mechanisms that contribute to global SLR (and their relative likelihood) as well as improved insights from paleoclimate studies (see section III-B). In summary, for the end of the 21st century, the most recent projections (i.e., those that incorporate ice sheet dynamics) indicate that global sea levels may rise by 0.7–1.0 m under RCP4.5 and 1.0–1.8 m under RCP8.5, and could even exceed 2 m (Bakker et al. 2017; Kopp et al. 2017; Le Bars et al. 2017; Wong et al. 2017; IPCC 2019).

Subsequent research by Bamber et al. (2019: page 11195) supports the Garner et al. (2018) findings by reiterating that "despite considerable advances in process understanding, numerical modeling, and the observational record of ice sheet contributions to global mean SLR since IPCC AR5, severe limitations remain in the predictive capability of ice sheet models. Consequently, the potential contributions of ice sheets remain the largest source of uncertainty in projecting future SLR." Bamber et al. (2019) report that, based on expert opinion, when inter- and intra-ice sheet processes and their tail dependences are accounted for, and thermal expansion and glacier contributions included, global SLR projections for 2100 that exceed 2.0 m are plausible (Table 5). This is consistent with United States National Oceanic and Atmospheric Administration (NOAA) (2017) guidelines, which (i) suggest that high levels of SLR (2.0–2.5 m) by 2100 must be considered plausible; and (ii) emphasize that SLR will not stop in 2100, so SLR scenarios greater than 2.0–2.5 m should be considered for projects with an expected lifetime beyond 2100. Note, however, that NOAA (2017) assigned 0.3% probability to SLR greater than 2.0 m by 2100 while Bamber et al. (2019) assess this probability to be greater than 5% (see 2100 H in Table 5).

Table 5: Global Mean Sea-Level Rise Projections

Centimeters above 2000 CE	50%	17–83%	5–95%	1–99%
2050 L	30	22–40	16–49	10–61
2050 H	34	26–47	21–61	16–77
2100 L	69	49–98	36–126	21–163
2100 L	111	79–174	62–238	43–329

CE = Common Era, H = High, L = Low.
Source: Bamber et al. (2019), Table 2.

The conclusions of Garner et al. (2018) and Bamber et al. (2019) and the resulting implications about the likelihood of significantly elevated global sea-level projections have not yet been assessed for the PIR. Nevertheless, both studies highlight the need for a more precautionary approach than simply adopting the upper global mean SLR projections when considering SLR impacts in the PIR. That is, higher-end scenarios should be considered as the latest science suggests that SLR greater than 1 m is conceivable at some point in the 21st century and SLR of 2 m by 2100 is plausible. It is also important to recognize that SLR will not stop in 2100. For example, Horton et al. (2018) estimate that the median global mean SLR between 2000 and 2150 is very likely to range from 0.3–1.5 m under RCP2.6, 0.4–3.2 m under RCP4.5, and 0.8–6.0 m under RCP8.5 (very likely ranges between 2000 and 2300 are −0.2–4.7 m under RCP2.6, 0.0–7.0 m under RCP4.5, and 1.0–15.5 m under RCP8.5).

C. Storm Surge Projections for the Pacific Islands Region

Although some global studies of future changes in storm surge exist (e.g., Mori et al. 2019; Muis et al. 2020), research for the PIR is limited. Most literature focuses on how the impacts of storm surge could be magnified in the future by superposition upon rising sea levels. Storm surges could change in frequency and/or intensity, mostly depending on if, how, where, and when the frequency of storms changes (McInnes et al. 2014, 2016a; Hoeke et al. 2015). Changes to interactions between storm surges, tides, and waves are also possible (McInnes et al. 2014; Mori et al. 2019; Muis et al. 2020). Given that storm surges are also determined by bathymetry and the shape of the coastline (in addition to the frequency, intensity, and track of storms), how storm surges in the PIR change in the future is highly site specific.

For example, modeling of tropical cyclone storm surges for Apia (Samoa) suggests that where a 1-in-50-year storm surge under baseline (1990) conditions would have caused only partial inundation of the western side of Mulinu'u Peninsula, a 1-in-100-year storm surge would completely inundate the peninsula (Hoeke et al. 2014). However, by 2055, increases in sea level could result in a 1-in-50-year storm surge, completely inundating the peninsula as well. Model results indicate that for a 1-in-100-year storm surge on top of future (2055) projected SLR, maximum sea levels on Mulinu'u Peninsula could be 2.6 m (midrange estimate) to 3.2 m (upper estimate) above current sea level.

Given the high degree of local variation in storm surge heights (and associated inundation), a more concerted effort is required to evaluate existing and future risks associated with storm surge in the PIR. A fundamental first step in this effort involves collating existing sources of bathymetry and topography, identifying where such data are absent or insufficient, and prioritizing efforts to gather data in missing areas. Water-level data across the PIR are required to validate modelled storm surge heights, which, in turn, would improve storm surge model accuracy and increase confidence in what the storm surge models say about storm surge heights under current and future conditions.

D.   Wave Climate Projections for the Pacific Islands Region

Wave climate is the description of wave characteristics (height, period, and direction) over time. Wave climate contributes to variability and change in sea levels via wave setup and wave run-up (section II-A). The PIR has been reported to be at least as vulnerable to variability and changes in wave climate as it is to increases in absolute sea level (Hemer et al. 2011; Wandres et al. 2020). It should also be acknowledged that submergence of protective features such as reefs because of SLR will lead to significant increases in shoreline wave heights without any changes in offshore wave heights, resulting in an increase in wave-driven flooding for most atolls (Storlazzi et al. 2018).

Wave climate is influenced by variability and/or changes in wind, particularly in regions, such as the PIR, affected by tropical cyclones (Church et al. 2013; Knutson et al. 2019, 2020). Hence, any variability or change in the intensity, frequency, duration, and/or path of tropical cyclones could alter the wave climate across the PIR, thereby modifying local sea levels. While the consensus is that anthropogenic climate change could impact tropical cyclone behavior, there is large uncertainty about how, when, and where tropical cyclone behavior in the PIR could change (Elsner et al. 2008; Knutson et al. 2010; Walsh et al. 2012; Emanuel 2013; Sugi et al. 2015; Walsh et al. 2016; Patricola and Wehner 2018; Knutson et al. 2019, 2020). Walsh (2015) projects a decrease in the number of tropical cyclones in the western (22%) and eastern (14%) South Pacific region by 2100. This is consistent with the projection of 17% decrease in the number of tropical cyclones (under RCP8.5 by 2100) by Bell et al. (2019) and could be related to an observed poleward shift in global tropical cyclone activity (Daloz and Camargo 2018, Sharmila and Walsh 2018). However, while the number of tropical cyclones in the PIR is projected to decrease, the intensity of the tropical cyclones that do occur could increase by 10%–20% (Parker et al. 2018, Patricola and Wehner 2018). Knutson et al. (2020) also have medium to high confidence that global average intensity of tropical cyclones may increase (with a median rise of ~5% for maximum surface wind speeds) and medium to high confidence that the proportion of tropical cyclones reaching very intense levels (categories 4 and 5) may increase.

Obstacles such as the brevity of a reliable tropical cyclone record (Magee and Verdon-Kidd 2019); nonhomogeneous environmental data (Sterl 2004); interannual variability of tropical cyclone activity

(Kuleshov et al. 2008; Magee et al. 2017); relatively coarse resolution of climate models (Henderson-Sellers et al. 1998; Walsh et al. 2016); and a lack of knowledge surrounding tropical cyclone formation, organization, and intensification (Walsh et al. 2016) all make it difficult to understand and model the impacts of future climate change on tropical cyclone activity in the PIR. These science gaps and challenges contribute to uncertainty about high-water levels for the PIR, since tropical cyclones strongly influence wave climate and wave climate is strongly associated with sea-level changes, but the location-specific implications for the PIR remain unclear (McInnes et al. 2014; Mori et al. 2019; Muis et al. 2020), highlighting the need for a more precautionary approach when considering SLR impacts in the PIR.

Dynamical models of wave climate suggest that significant wave height may increase (decrease) by up to 0.2 m for the eastern (western) equatorial Pacific region during austral winter (Hemer et al. 2011), with greater decreases to the north of the equator (Trenham et al. 2013). Increases in annual mean wave periods by the end of the 21st century were also suggested by Hemer et al. (2011) but Trenham et al. (2013) found no statistically significant change in the mean wave period. However, Trenham et al. (2013) did project changes in wave direction in the austral winter, namely, an increased southerly component due to projected increases in Southern Ocean storminess and an enhanced easterly component associated with projected increased strength of the easterly trade winds.

The PACCSAP https://www.pacificclimatechangescience.org/ program produced regionally specific projections for changes to wave height, wave period, and wave direction for 15 countries and territories across the PIR (Australian Bureau of Meteorology and Commonwealth Scientific and Industrial Research Organisation 2014). Appendix 2 shows the location-specific changes in wave height, wave period, and wave direction in the PIR for RCP4.5 and RCP8.5 for the 20 years centered on 2035 and 2090). Consistent with more recent literature (e.g., Mori et al. 2019; Muis et al. 2020), in all PIR locations covered in Australian Bureau of Meteorology and Commonwealth Scientific and Industrial Research Organisation (2014), there is only low confidence associated with the projected changes in wave height, wave period, and wave direction. This uncertainty again highlights the need for a more precautionary approach when considering SLR impacts in the PIR.

E.   Wave Power Projections for the Pacific Islands Region

Until recently, most analyses of wave climate focused on historical trends and future projections of mean and extreme values for parameters such as wave height, wave period, and wave direction (section IV-D). However, Reguero et al. (2019) showed that global wave power, which is the transfer of energy from the wind into sea-surface motion, increased globally by 0.47% per year between 1948 and 2008 and by 2.3% per year from 1994. Wave power in the Southern Ocean (defined by the 40°S latitudinal limit) has increased by 0.58% per year, while wave power in the Pacific has increased by 0.35% per year compared with 0.4% per year increases in the Atlantic Ocean and Indian Ocean. These trends are statistically significant and caused by the influence of sea surface temperatures (SSTs) on wind patterns (Reguero et al. 2019). However, it is important to note that Reguero et al. (2019) used a wave hindcast forced with Climate Forecast System Reanalysis (CFSR) data. CFSR is poorly suited for assessing historical trends because of homogeneity issues, plus CFSR data are not well aligned with satellite altimeter observations or other reanalysis products. Young and Ribal (2019) raised potential issues with the Reguero et al. (2019) conclusions when they analyzed global satellite data from 1985 to 2018 and found only small increases in oceanic wind speed and wave height. They found small increases in both quantities, with the strongest increases in extreme conditions and in the Southern Ocean. Timmermans et al. (2020) further highlighted the uncertainties associated with historical and future wave power estimates when they analyzed trends using two products based on satellite altimetry and two reanalysis hindcast datasets. They found general similarity in spatial variation and magnitude but major differences in equatorial regions and the Indian Ocean, raising fundamental questions about uncertainty in all data products used to assess wave power. Therefore, further research is required to better understand the past, present, and future impact of wave power on wave setup and wave run-up (section II-A) and the resulting impact on high-water levels in the PIR.

V. Uncertainties and Limitations Associated with the Science on Sea-Level Rise in the Pacific Islands Region

This review of evidence about sea levels for the PIR reveals that substantial changes have occurred in the past and more changes are expected in the future. Although studies agree that sea levels in the PIR are rising and will continue to do so, there is large uncertainty about the magnitude, timing and location of the projected SLR. Even larger uncertainties exist around how high-water levels could be influenced (i.e., amplified or moderated) by natural climate variability, storm surges, and changes to wave climate and wave power.

The most important uncertainties and knowledge gaps identified by this review include the following:

(i) **Limited historical data and gaps in reporting networks.** To credibly investigate the future, we must understand the past. Across the PIR, limited observational hydrometeorological and coastal water level datasets are available. The lack of systematic measurements and low density of in-situ wave and water-level monitoring in wave-exposed regions mean studies are reliant on tide-gauge data. Tide-gauge data do not provide a comprehensive understanding of risk caused by remotely generated swell because tide gauges are typically sited in sheltered locations and are not designed to measure waves (Hoeke et al. 2013; McInnes et al. 2016b). Although a few wave buoys exist, records are typically intermittent and short-lived (Trenham et al. 2013). A strategic, long-term wave monitoring and observation program would enable enhanced coastal hazard assessments (Hemer et al. 2011).

(ii) **Range of SLR projections.** Variations between model outputs suggest that some aspects of sea level (both globally and regionally) remain poorly understood. Church et al. (2013) called for improved parameterizations of unresolved physical processes, improved numerical algorithms, and finer grid resolutions to better simulate features such as boundary currents and mesoscale eddies. In addition, changes in future greenhouse gas emissions are uncertain. Further uncertainties surround the magnitude and rate of the ice sheet contribution to SLR and the regional distribution of SLR (Church et al. 2013). The IPCC Special Report on the Oceans and Cryosphere in a Changing Climate (SROCC) (2019) and AR6 (IPCC 2021) addressed some of these issues. However, more work is needed to better resolve uncertainties surrounding SLR, especially upper-bound projections and abrupt-change scenarios. In the meantime, a more precautionary approach than simply adopting the upper global mean SLR projections is recommended when considering SLR impacts in the PIR. That is, higher-end scenarios must be included as the latest science suggests the SLR scenarios discussed in section IV-B could occur at some point in the 21st century.

(iii) **Relative contribution of tides, waves, storm surge, and bathymetry (and their interactions) to sea levels across the PIR.** Information about the relative contribution of tides, storm surge, wave climate, and wave power in amplifying sea levels across the PIR is limited. Without a regionally specific understanding of the role these processes play in different parts of the PIR, it is difficult to understand how future climate change could affect sea levels at specific locations in the PIR (Walsh et al. 2012; Hoeke et al. 2013; Winter et al. 2020).

(iv) **Location-specific variations.** Small islands do not have uniform climate change risk profiles (Nurse et al. 2014; Winter et al. 2020) and some small island states span vast areas of ocean. Island nations are composed of different geomorphological units and their locations determine the relative influence of climate variability and change. This includes impacts associated with, for example, intraseasonal, interannual, and/or interdecadal climate variability and tropical cyclones. More work is needed to produce location-specific projections for absolute SLR, storm surge, wave climate, and wave power across the PIR. Regionally specific projections that incorporate the AR6 science and SLR modeling could assist decision makers, environmental agencies, and PIR governments to make more informed decisions about climate adaptation and disaster risk reduction measures (Aucan 2018). Work is underway to update the PACCSAP program location-specific SLR projections for the PIR (Figure 11, Figure 12, Table 4, and Appendix 1) to incorporate the AR6 science and SLR modeling. In the meantime, localized versions of the AR6 projections can also be viewed at the NASA and IPCC Sea Level Projection Tool https://sealevel.nasa.gov/ipcc-ar6-sea-level-projection-tool.

(v)   **Changes to future tropical cyclone behavior.** Changes in the frequency, location, and duration of future tropical cyclones could have significant implications for high-water levels across the PIR. More work is needed to (a) quantify whether, how, where, and when future tropical cyclone activity could change; and (b) evaluate how projected changes in tropical cyclone behavior translate into storm surge, wave climate, and wave power, and how, where, and when these factors amplify changes in absolute sea levels.

(vi)   **Future of interannual and/or interdecadal modes of variability.** ENSO and IPO are known to drive significant variations in sea levels across the PIR. Although future sea level "seesaws" in the tropical Pacific are expected to continue, it is unclear whether or how ENSO and/or IPO variability might change and what this could mean for PIR nations and territories.

(vii)   **Vertical land movement.** Relative to SLR, vertical land movement is an important consideration that can amplify or offset absolute SLR (e.g., Becker et al. 2012; Ballu et al. 2011; Aucan 2018). Records of vertical land movement in different locations across the PIR are typically short and many are not continuous (section II-C). Given the importance of vertical land movement, more effort is needed to install and maintain monitors of vertical land movement to compile reliable, detailed, long-term records of vertical land movement. This would support more robust risk assessment and management.

(viii)   **Compound extreme events.** This report does not consider potential changes to the frequency, location, or sequencing of extreme rainfall events, or the impacts associated with compound extreme events (e.g., SLR in addition to increased rainfall frequency or intensity, pluvial, and fluvial flooding). See, for example, Moftakhari et al. (2017) for further details.

(ix)   **Errors and/or bias in digital elevation models (DEMs).** Land topography and elevation, as represented by DEMs, are used to translate SLR observations and projections into socioeconomic and environmental impacts (e.g., by coastal inundation mapping, population exposure assessments, among others). A typical choice for assessing exposure to SLR is the 30 m horizontal resolution NASA Shuttle Radar Topography Mission (SRTM) https://www2.jpl.nasa.gov/srtm/. Kulp and Strauss (2019) showed that SRTM has a global mean positive vertical bias of ~2 m, which is comparable to global mean SLR estimates for 2100 and suggests that vulnerability to SLR could be significantly underestimated where SRTM is used as the DEM and where the SRTM bias is positive. Other DEMs are associated with bias and uncertainties. To address these DEM-related uncertainties, Kulp and Strauss (2019: page 8) called for "development and public release of improved coastal area elevation datasets building directly off of new high-resolution observations increasingly collected by satellites today." A new altimeter system called Ice, Cloud and Land Elevation Satellite (ICESat-2) https://icesat-2.gsfc.nasa.gov/ uses lasers and a precise detection instrument to measure the elevation of Earth's surface. Research is required to assess the data provided by ICESat-2 to see whether they reduce errors and/or bias in DEMs and result in improved SLR-related observations and impact assessments. Research should also be conducted into the usefulness of the first global elevation model derived from satellite light detection and ranging (LiDAR) data (Hooijer and Vernimmen 2021) for PIR SLR impact assessments. Development and/or use of other LiDAR information for key locations across the PIR should be investigated.

(x)   **Implications of the gravitational fingerprint for the PIR.** Ice sheets and glaciers have a gravitational pull on the water that surrounds them, making the sea level a little higher at their edges. When a glacier or ice sheet melts, it loses mass; therefore, the gravitational pull exerted on nearby ocean water weakens and the sea level falls. At the same time, the land rises because the ice is no longer weighing it down, resulting in a further drop in relative sea level. The redistribution of mass changes the Earth's gravitational field, causing the fresh meltwater and ocean water to move away toward remote coastlines; the resulting pattern of SLR is the fingerprint of melting from that particular ice sheet or glacier (Bamber and Riva 2010; Kopp et al. 2014; Hsu and Velicogna 2017). Hence, the gravitational fingerprint for the PIR will depend on the combined patterns of future melt from the Greenland and Antarctic ice sheets. For Hawaii, recent work suggests that for 1 m of SLR, the water level rises by as much as 20% because of the gravitational fingerprint: the greater the SLR, the greater the multiplier because of the gravitational fingerprint (NOAA 2017, Figure 7; Sweet at al. 2017, section 12.2, Figure 12.1). It is uncertain whether something similar applies for different locations across the PIR.

(xi)   **Changes in tidal range and/or position of the amphidromic points with SLR.** An amphidromic point, also called a tidal node, is a geographical location that has zero tidal amplitude for one harmonic constituent of the tide. The tidal range (peak-to-peak amplitude or height difference between high and low tide) for that harmonic constituent increases with distance from the point. If and/or how amphidromic points could change with SLR and the impacts this could have in the PIR are uncertain (see Pickering et al. (2017) and Haigh et al. (2020) for further details).

VI. Advice on Developing Guidance on How to Incorporate Credible Sea-Level Rise Information into ADB Projects

SLR projections from AR5 suggested that SLR in the PIR is unlikely to exceed 1 m by 2100 (relative to the 1986–2005 baseline used in AR5). This information, along with knowledge that most projects have service lifetimes of less than 80 years, has been widely used by ADB and others to assess and manage SLR-related risks in the PIR, with allowance and adaptation for SLR of up to 1 m considered sufficiently precautionary for projects with operational lifetimes of less than 100 years. However, the key message from this review of science and evidence that has emerged since AR5 was published is that there are several reasons why such an approach may not always be adequate in the PIR:

(i) Figure 11 and Figure 12 (section IV-A) and Appendix 1 indicate that although there is very high confidence in the direction of change for all PIR locations (i.e., decrease in sea level is not projected anywhere in the PIR), there is only medium confidence in the magnitude and timing of change. This is due to uncertainties associated with (a) projections of the Greenland and Antarctic ice sheet contributions (Slater et al. 2021); (b) the influence of natural interannual to decadal variability, which could lead to conditions where sea levels are further elevated (e.g., because of increased tropical cyclones [see page 4, second paragraph] or increased La Niña [see page 6, second paragraph]); and (c) the gravitational fingerprint associated with redistribution of water from Greenland and Antarctic ice melt (Bamber and Riva 2010; Hsu and Velicogna 2017). Nevertheless, for most locations in the PIR where location-specific analysis has been conducted and where the impacts of natural climate variability are considered, it is possible that SLR might exceed 1 m by the end of the 21st century (relative to the 1986–2005 baseline used in AR5).

(ii) AR6 and other work that has emerged since AR5 demonstrate that global SLR greater than 1 m by the end of the 21st century is conceivable. One expert elicitation suggests a greater than 5% probability that SLR would exceed 2 m by 2100 (Bamber et al. 2019), although other literature (NOAA 2017, IPCC 2019) assigns a lower probability (0.3%) to the scenario of 2 m SLR by 2100.

(iii) Some paleoclimate records suggest that SLR of 5 m in a century has occurred before. However, the consensus view is that such extreme SLR would happen over long periods (centuries to millennia) and is unlikely to occur before 2100. AR6 states that projected global mean SLR of 1.7–6.8 m by 2300 is possible without marine ice cliff instability and this projection increases to 16 m by 2300 with marine ice cliff instability.

(iv) Short-term variability in high-water levels associated with storm surge and waves could significantly increase local coastal water levels above what is expected because of changes to absolute sea level (i.e., changes to long-term average sea level alone), especially in the PIR and particularly in the western tropical Pacific because of the high exposure to tropical cyclones, high shoreline–land area ratio, high exposure to waves and currents, combined with low-lying coral atolls, reefs, or volcanically composed islands.

(v) Based on observed data collected since ~2000, most islands in the PIR are evidently subsiding (i.e., have negative vertical land movement). Therefore, irrespective of any other influence, the effect of SLR will be magnified where land is falling, which appears to be the case for much of the PIR.

The implications of the potential for significantly elevated global SLR projections for the PIR have not yet been assessed, and it is not feasible to do the detailed modeling work required to rigorously quantify location-specific SLR-related risks for every ADB project in the region. However, when considering all the factors that contribute to SLR impacts (e.g., SLR, storm surge; wave setup, run-up, and power; and vertical land movement, among others) (Quataert et al. 2015; Vitousek et al. 2017), the findings of this review highlight the need for a more precautionary approach than simply adopting the upper global mean SLR projections when considering the impacts of high-water levels in climate risk and adaptation assessments (CRAs) for the PIR. There is a need to consider both the likely range of SLR and higher-end scenarios should be considered as the latest science suggests that SLR greater than 1 m is conceivable at some point in the 21st century and that SLR of 2 m by 2100 is plausible. It is also important to recognize that SLR will not stop in 2100 and using 2100 as a planning time frame is arbitrary; a 100-year planning time frame (i.e., 2122) might be more appropriate for long-term planning.

Therefore, in line with Hinkel et al. (2019) and Stammer et al. (2019), it is advised that a precautionary approach for ADB CRAs in the PIR requires the following to be considered relative to the 1995–2014 baseline introduced in AR6:

(i)   for all projects, a 1 m SLR scenario, for comparison with studies that have typically used a scenario of 1 m SLR by 2100;

(ii)   for short- to medium-term projects (i.e., with a design life of 20–30 years), a scenario of 0.5 m SLR by 2050;

(iii)   for long-term projects (i.e., with a design life greater than 30 years), a scenario of 2 m SLR by 2100; and

(iv)   for projects with an expected lifetime beyond 2100, scenarios greater than 2 m SLR.

These SLR scenarios should be used not only in sensitivity analyses of climate proofing design and options but also in the analysis of the costs and benefits of additional climate proofing, to explore the flexibility of adaptation options (and to avoid maladaptation). The scenarios should also draw on ADB (2015) guidance on the economics of climate proofing. It is emphasized that these SLR scenarios are recommended for sensitivity analysis rather than as minimum precautionary levels for climate proofing. The flexibility provided by adaptive management approaches could also address higher SLR, noting this needs to consider the lifetime, risk of lock-in, and level of precaution associated with investments. Where warranted (i.e., at sites with high exposure and/or vulnerability), extra allowance should also be made for the influence of natural climate variability, tropical cyclones, storm surge, waves, and vertical land movement. Exactly what that allowance should be will depend on the type of project and its location in the PIR, as well as the appetite for risk and expected lifetime of the project. Path dependency should also be considered. That is, decisions to invest (or otherwise) in coastal infrastructure based on assumptions about possible high-water levels and other factors, including economics, will influence subsequent investments and development, for which the same level of risk assessment may not be performed. Refer to PRIF (2021) for further guidance.

The ADB guidance document to be developed based on the findings of this review will follow the latest thinking on robust decision-making under uncertainty and on integration of SLR information into coastal risk and adaptation assessments (e.g., adaptation pathways, real options, flexible decision-making, avoidance of costly or overengineered upfront adaptation options) (NCCARF 2016; NOAA 2017; Hinkel et al. 2019; Stammer et al. 2019; Nicholls et al. 2021). The ADB guidance document will include worked examples, similar to the approach taken by Wilby et al. (2011, 2014), for some case study locations in the PIR to identify explicitly how different components of SLR are considered and to identify critical location-specific thresholds and existing and/or potential local planning and/or adaptation options. These worked examples will also explain how to assess and manage SLR risks in data-poor and data-rich environments (Duong et al. 2017, 2018).

VII. Recommendations for Future Work

(i) **Hold a workshop to discuss and scope future tasks.** Conduct a workshop involving relevant experts to (i) discuss this report (as well as the SLR calculator and associated knowledge product that ADB has already developed for Viet Nam) (ADB 2020a, 2020b), which will assist PARD in developing supplementary guidance on incorporating credible SLR projection information on climate risk management into ADB projects; (ii) prioritize the recommendations and tasks emerging from the workshop to develop supplementary guidance on assessing SLR in the PIR; and (iii) plan the projects and terms of reference needed to address the specific technical recommendations and tasks emerging from this report and the workshop. Required skills and personnel must be identified and a realistic budget and timeline estimated for each piece of work.

(ii) **Develop Pacific Islands Region case studies to demonstrate the use of sea-level rise scenarios.** This could involve applying this report's advice to existing ADB projects in the PIR to see how it affects the technical design and project economics and to provide good-practice exemplars. The case studies should include not only physical analysis and engineering aspects but also economics. Conducting case studies before the workshop (Recommendation #1) might be useful so that challenges encountered can be discussed and tackled by the experts present at the workshop.

(iii) **Conduct a sensitivity analysis to determine what increased sea-level rise estimates could mean for existing, long-lived ADB projects in the Pacific Islands Region.** Sensitivity analysis is required for such projects to assess the impacts of allowing for up to 2 m of SLR in the 21st century compared with the current practice of allowing for 1 m by 2100. This due diligence should be incorporated into CRAs as another plausible scenario to inform existing and/or proposed adaptation measures. A first step in the due diligence exercise is to determine how many existing, long-lived ADB projects in the PIR need to be revisited.

(iv) **Improve sources and coverage of bathymetry, topography, and water-level data.** To better understand the impacts of storm surge and how it exacerbates the SLR impacts (section IV-C), an intensive effort should be made to collate sources of bathymetry and topography, identify where such data are absent or insufficient, and prioritize efforts to collect data in areas with no or insufficient data. High-resolution water-level data across the PIR are also required to validate modelled storm surge heights, which, in turn, could improve model accuracy and increase confidence in what the storm surge models say about the future. See Winter et al. (2020) for further perspectives on the issue from an international group of scientists. The task is potentially huge and costly and a scoping study of how best to do it would be a logical first step.

(v) **Improve sources and coverage of wave climate and wave power data.** Changes in wave climate and wave power are anticipated, but information is scarce about whether, how, and where the PIR could be affected (sections IV-D and IV-E). Work, mostly desktop, is needed to establish what data exist (either observed or modelled) for wave climate and wave power in the PIR. Where enough data exist, analysis should be done on whether, how, and where wave climate and wave power have changed or are projected to change for specific locations. Where data gaps exist, networks monitoring wave climate and wave power should be upgraded. Insights emerging from this work could provide information about the impact of wave power on wave setup and wave run-up and the resulting impact on sea-level changes in the PIR. See Winter et al. (2020) for further perspectives on this issue from an international group of scientists.

(vi) **Upgrade altimetry and digital elevation models in the Pacific Islands Region.** Vertical land movement is important when considering SLR impacts, but fewer than 20 inhabited islands in the PIR have systems in place to monitor land movement (section II-C). Improvements in the resolution, coverage, and quality of land elevation information are necessary to (i) better understand and quantify land movement and (ii) undertake detailed geographic information system (GIS)–based assessments of SLR impacts. Availability of sediment for vertical island accretion must be assessed and monitored, particularly in the context of additional stress on coral systems and whether the availability of sediment can be maintained (Masselink et al. 2020).

(vii) **Improve understanding and quantification of the causes and impacts of short-term variability in high-water levels.** The impacts of natural climate variability (e.g., ENSO, IPO) and tropical cyclones on historical sea levels and storm surge events at specific project sites need to be better understood. It is essential for establishing reliable estimates of baseline SLR to which SLR projections and vertical land movement are applied. If the baseline SLR and associated risk estimates do not properly account for the impacts of natural climate variability, then the feasibility, sensitivity, and cost–benefit assessments may be flawed even if precautionary SLR scenarios of greater than 2 m are used.

(viii) **Periodically review sea-level rise evidence, science, and adaptation.** The science on SLR is evolving rapidly and views are changing about how best to deal with existing and projected SLR impacts. Adaptation to the impacts of climate change (including SLR) is also an ongoing process. The evidence should be reviewed every 5–10 years, ideally as soon as possible after the release of each new IPCC assessment report, to determine whether risks have changed in light of new evidence and, if required, to update the recommendations and guidance so that ADB activities are consistent with the best science and practice. The suitability and robustness of SLR adaptation strategies should also be reviewed every 5–10 years to either confirm their appropriateness or, if needed, implement different actions on the adaptation pathway.

Appendixes

Appendix 1: Location-Specific Changes in Sea Level in the Pacific Islands Region

		2030	2050	2070	2090	Confidence in Projected Magnitude of Change
Cook Islands (northern)	RCP2.6	12 (7–17)	21 (13–29)	30 (18–43)	39 (22–57)	Medium
	RCP4.5	12 (7–17)	22 (14–30)	33 (21–47)	46 (28–65)	
	RCP6	11 (7–16)	21 (13–29)	33 (20–46)	46 (28–66)	
	RCP8.5	12 (8–17)	24 (16–33)	40 (26–56)	61 (39–86)	
Cook Islands (southern)	RCP2.6	12 (7–17)	21 (13–29)	30 (18–43)	39 (22–57)	Medium
	RCP4.5	12 (7–17)	22 (14–30)	33 (21–47)	46 (28–65)	
	RCP6	11 (7–16)	21 (13–29)	33 (20–46)	46 (28–66)	
	RCP8.5	12 (8–17)	24 (16–33)	40 (26–56)	61 (39–86)	
Federated States of Micronesia (east)	RCP2.6	13 (8–18)	22 (14–30)	32 (20–45)	42 (24–60)	Medium
	RCP4.5	12 (8–17)	22 (14–31)	35 (22–49)	48 (30–68)	
	RCP6	12 (7–17)	22 (14–30)	34 (22–48)	49 (31–69)	
	RCP8.5	13 (8–18)	26 (17–35)	43 (28–59)	64 (41–90)	
Federated States of Micronesia (west)	RCP2.6	13 (8–18)	22 (14–30)	32 (20–45)	42 (24–60)	Medium
	RCP4.5	12 (8–17)	22 (14–31)	35 (22–49)	48 (30–68)	
	RCP6	12 (7–17)	22 (14–30)	34 (22–48)	49 (31–69)	
	RCP8.5	13 (8–18)	26 (17–35)	43 (28–59)	64 (41–90)	
Fiji	RCP2.6	13 (8–18)	22 (14–31)	31 (19–44)	41 (24–58)	Medium
	RCP4.5	13 (8–18)	23 (14–31)	35 (22–48)	47 (29–67)	
	RCP6	13 (8–17)	22 (14–31)	34 (22–47)	49 (30–68)	
	RCP8.5	13 (8–18)	25 (17–35)	42 (28–58)	64 (41–88)	
Kiribati (Phoenix Group)	RCP2.6	12 (7–17)	21 (13–29)	31 (18–44)	40 (23–59)	Medium
	RCP4.5	12 (7–16)	22 (13–30)	33 (20–47)	46 (27–66)	
	RCP6	11 (7–16)	21 (13–29)	33 (19–46)	47 (28–67)	
	RCP8.5	12 (7–17)	24 (16–33)	40 (26–56)	61 (38–87)	
Kiribati (Line Group)	RCP2.6	12 (7–17)	21 (13–29)	31 (18–44)	40 (23–59)	Medium
	RCP4.5	12 (7–16)	22 (13–30)	33 (20–47)	46 (27–66)	
	RCP6	11 (7–16)	21 (13–29)	33 (19–46)	47 (28–67)	
	RCP8.5	12 (7–17)	24 (16–33)	40 (26–56)	61 (38–87)	

continued on next page

Appendix 1 continued

		2030	2050	2070	2090	Confidence in Projected Magnitude of Change
Marshall Islands (northern)	RCP2.6	13 (7–18)	22 (13–30)	31 (19–45)	41 (23–60)	Medium
	RCP4.5	12 (7–18)	23 (14–32)	35 (21–49)	48 (28–69)	
	RCP6	12 (7–17)	22 (14–31)	35 (21–49)	49 (30–70)	
	RCP8.5	13 (8–19)	26 (16–35)	43 (27–60)	65 (41–92)	
Marshall Islands (southern)	RCP2.6	13 (7–18)	22 (13–30)	31 (19–45)	41 (23–60)	Medium
	RCP4.5	12 (7–18)	23 (14–32)	35 (21–49)	48 (28–69)	
	RCP6	12 (7–17)	22 (14–31)	35 (21–49)	49 (30–70)	
	RCP8.5	13 (8–19)	26 (16–35)	43 (27–60)	65 (41–92)	
Nauru	RCP2.6	12 (8–17)	22 (14–30)	32 (19–45)	42 (24–60)	Medium
	RCP4.5	12 (7–17)	22 (14–31)	35 (22–48)	48 (29–68)	
	RCP6	12 (7–16)	22 (14–30)	34 (21–48)	49 (30–69)	
	RCP8.5	13 (8–18)	25 (17–34)	42 (28–58)	63 (41–89)	
Niue	RCP2.6	12 (7–17)	21 (13–30)	30 (18–43)	40 (23–57)	Medium
	RCP4.5	12 (8–17)	22 (14–31)	34 (22–47)	47 (29–66)	
	RCP6	12 (7–17)	22 (13–30)	34 (21–47)	47 (29–67)	
	RCP8.5	13 (8–18)	25 (16–34)	42 (27–57)	62 (41–87)	
Palau	RCP2.6	13 (8–17)	22 (14–30)	32 (20–44)	42 (25–59)	Medium
	RCP4.5	12 (8–17)	23 (15–31)	35 (23–48)	48 (30–67)	
	RCP6	12 (8–17)	22 (14–30)	34 (22–47)	49 (31–68)	
	RCP8.5	13 (8–18)	26 (17–35)	43 (28–58)	64 (41–88)	
Papua New Guinea	RCP2.6	12 (8–17)	22 (14–30)	31 (19–44)	41 (24–58)	Medium
	RCP4.5	12 (7–17)	22 (14–31)	34 (22–47)	47 (29–66)	
	RCP6	12 (7–16)	22 (14–29)	34 (21–46)	48 (30–67)	
	RCP8.5	13 (8–17)	25 (17–34)	42 (28–57)	63 (47–87)	
Samoa	RCP2.6	12 (8–17)	21 (13–30)	31 (18–44)	41 (23–59)	Medium
	RCP4.5	12 (7–17)	22 (13–30)	34 (21–47)	46 (28–66)	
	RCP6	12 (7–17)	21 (13–29)	33 (21–46)	48 (29–67)	
	RCP8.5	12 (7–17)	24 (16–33)	41 (27–56)	62 (40–87)	
Solomon Islands	RCP2.6	13 (8–18)	22 (14–31)	32 (19–45)	42 (24–60)	Medium
	RCP4.5	12 (7–17)	22 (14–31)	35 (21–48)	47 (29–67)	
	RCP6	12 (7–17)	22 (14–30)	34 (21–47)	49 (30–69)	
	RCP8.5	13 (8–18)	25 (16–35)	42 (28–58)	63 (40–89)	
Tonga	RCP2.6	13 (8–18)	22 (14–30)	31 (19–43)	40 (23–58)	Medium
	RCP4.5	13 (8–18)	23 (15–31)	35 (22–48)	47 (29–66)	
	RCP6	12 (7–17)	22 (14–31)	34 (21–47)	48 (30–67)	
	RCP8.5	13 (8–18)	25 (17–35)	42 (28–58)	63 (41–88)	

continued on next page

Appendix 1 continued

		2030	2050	2070	2090	Confidence in Projected Magnitude of Change
Tuvalu	RCP2.6	12 (7–17)	21 (13–30)	31 (19–44)	41 (23–59)	Medium
	RCP4.5	12 (7–17)	22 (13–31)	34 (20–48)	47 (28–67)	
	RCP6	12 (7–16)	21 (13–29)	33 (20–47)	48 (28–67)	
	RCP8.5	12 (7–18)	24 (16–34)	41 (26–57)	62 (39–87)	
Vanuatu	RCP2.6	13 (8–19)	23 (15–31)	32 (20–45)	42 (25–59)	Medium
	RCP4.5	13 (8–18)	23 (15–32)	36 (23–49)	48 (30–67)	
	RCP6	13 (8–18)	23 (15–31)	35 (23–48)	50 (32–69)	
	RCP8.5	13 (8–18)	26 (17–35)	43 (29–59)	64 (42–89)	

RCP = Representative Concentration Pathway.

Notes: The table shows projected changes in annual mean sea level for 15 island nations and territories across the Pacific Islands Region under four emission scenarios (RCP2.6, RCP4.5, RCP6.0, RCP8.5). Projected changes are given for four 20-year periods centered on 2030, 2050, 2070 and 2090, relative to a 20-year period centered on 1995. Values represent the multimodal mean change, with the 5%–95% range of uncertainty in brackets. Confidence in magnitude of change is expressed as high, medium, or low.

Source: Australian Bureau of Meteorology and Commonwealth Scientific and Industrial Research Organisation (2014).

Appendix 2: Location-Specific Changes in Wave Height, Wave Period, and Wave Direction in the Pacific Islands Region

	Variable	Season	2035	2090	Confidence in Projected Changes
Cook Islands (Northern)	Wave height change (m)	December–March	−0.0 (−0.3–0.2) −0.0 (−0.3–0.2)	−0.1 (−0.3–0.2) −0.1 (−0.3–0.2)	Low
		June–September	0.0 (−0.2–0.2) +0.0 (−0.1–0.2)	(−0.1–0.2) +0.0 (−0.2–0.3)	
	Wave period change (s)	December–March	+0.0 (−1.7–1.8) −0.0 (−1.7–1.6)	−0.1 (−2.0–1.9) −0.1 (−2.2–2.0)	
		June–September	+0.0 (−1.1–1.2) +0.0 (−1.1 to1.1)	−0.0 (−1.3–1.2) −0.0 (−1.4–1.3)	
	Wave direction change (° clockwise)	December–March	+0 (−30–40) 0 (−30–30)	0 (−30–30) 0 (−40–40)	
		June–September	0 (−10–10) 0 (−10–10)	−0 (−20–10) −0 (−20–10)	
Cook Islands (Southern)	Wave height change (m)	December–March	0.0 (−0.3–0.2) −0.0 (−0.3–0.2)	−0.0 (−0.3–0.2) −0.1 (−0.3–0.2)	Low
		June–September	+0.0 (−0.3–0.4) 0.0 (−0.3–0.3)	0.0 (−0.4–0.4) 0.0 (−0.4–0.4)	
	Wave period change (s)	December–March	−0.0 (−1.5–1.4) −0.0 (−1.4–1.3)	−0.1 (−1.7–1.5) −0.1 (−1.9–1.6)	
		June–September	0.0 (−0.9–0.9) 0.0 (−0.9–0.9)	0.0 (−1.1–1.1) −0.0 (−1.2–1.1)	
	Wave direction change (° clockwise)	December–March	0 (−70–70) 0 (−60–60)	−0 (−60–60) −10 (−70–60)	
		June–September	−0 (−20–10) −0 (−20–10)	−0 (−20–10) −10 (−20–5)	
Federated States of Micronesia (East)	Wave height change (m)	December–March	−0.0 (−0.2–0.2) −0.1 (−0.3–0.2)	−0.1 (−0.3–0.1) −0.2 (−0.4––0.0)	Low
		June–September	+0.0 (−0.2–0.2) +0.0 (−0.1–0.2)	0.0 (−0.2–0.2) +0.0 (−0.1–0.2)	
	Wave period change (s)	December–March	−0.1 (−0.6–0.4) −0.1 (−0.6–0.5)	−0.1 (−0.7–0.5) −0.2 (−0.9–0.4)	
		June–September	0.0 (−0.6–0.6) −0.0 (−0.6–0.5)	−0.0 (−0.7–0.6) −0.1 (−0.7–0.5)	
	Wave direction change (° clockwise)	December–March	0 (−10–10) 0 (−10–10)	0 (−10–10) 0 (−10–10)	
		June–September	+0 (−20–40) 0 (−20–40)	+0 (−20–30) +10 (−20–60)	

continued on next page

Appendix 2 continued

	Variable	Season	2035	2090	Confidence in Projected Changes
Federated States of Micronesia (West)	Wave height change (m)	December–March	−0.0 (−0.3–0.2) −0.0 (−0.3–0.2)	−0.1 (−0.3–0.1) −0.2 (−0.4–0.0)	Low
		June–September	0.0 (−0.2–0.2) 0.0 (−0.2–0.2)	0.0 (−0.2–0.2) 0.0 (−0.2–0.2)	
	Wave period change (s)	December–March	−0.1 (−0.4–0.3) −0.1 (−0.5–0.3)	−0.1 (−0.5–0.3) −0.3 (−0.7–0.2)	
		June–September	0.0 (−0.4–0.4) 0.0 (−0.4–0.4)	−0.0 (−0.5–0.4) −0.1 (−0.6–0.4)	
	Wave direction change (° clockwise)	December–March	0 (−5–5) 0 (−5–5)	0 (−10–5) −0 (−10–5)	
		June–September	+0 (−40–40) 0 (−40–40)	+0 (−30–40) +10 (−30–50)	
Fiji	Wave height change (m)	December–March	0.0 (−0.2–0.2) −0.0 (−0.2–0.2)	−0.0 (−0.3–0.2) −0 1 (−0.3–0.1)	Low
		June–September	+0.0 (−0.3–0.4) +0.0 (−0.3–0.3)	+0.0 (−0.3–0.4) +0.0 (−0.3–0.4)	
	Wave period change (s)	December–March	−0.1 (−0.9–0.7) −0.0 (−0.9–0.8)	−0.1 (−1.0–0.8) −0.1 (−1.1–0.9)	
		June–September	+0.0 (−0.9–0.9) +0.0 (−0.8–0.9)	+0.1 (−1.0–1.2) +0.1 (−1.1–1.2)	
	Wave direction change (° clockwise)	December–March	0 (−20–20) 0 (−20–20)	0 (−20–20) 10 (−20–30)	
		June–September	0 (−10–10) 0 (−10–10)	0 (−10–10) −0 (−10–10)	
Kiribati (Gilbert Islands)	Wave height change (m)	December–March	−0.0 (−0.2–0.1) −0.1 (−0.2–0.1)	−0.1 (−0.2–0.1) −0.2 (−0.3–0.1)	Low
		June–September	0.0 (−0.1–0.1) 0.0 (−0.1–0.1)	0.0 (−0.1–0.1) 0.0 (−0.1–0.1)	
	Wave period change (s)	December–March	−0.0 (−1.0–1.5) −0.1 (−1.3–1.1)	−0.1 (−1.2–1.6) −0.2 (−1.3–1.5)	
		June–September	+0.1 (−0.5–0.7) 0.0 (−0.6–0.6)	+0.0 (−0.7–0.7) −0.0 (−0.7–0.7)	
	Wave direction change (° clockwise)	December–March	0 (−10–10) 0 (−10–10)	0 (−10–10) 0 (−10–10)	
		June–September	+0 (−10–10) 0 (−10–10)	+0 (−10–20) +0 (−10–20)	

continued on next page

Appendix 2 continued

	Variable	Season	2035	2090	Confidence in Projected Changes
Kiribati (Phoenix Islands)	Wave height change (m)	December–March	−0.0 (−0.3–0.2) −0.0 (−0.3–0.2)	−0.1 (−0.3–0.2) −0.1 (−0.4–0.2)	Low
		June–September	0.0 (−0.1–0.2) 0.0 (−0.1–0.1)	0.0 (−0.1–0.2) +0.0 (−0.2–0.2)	
	Wave period change (s)	December–March	0.0 (−1.5–1.8) −0.1 (−1.4–1.6)	−0.1 (−1.8–2.0) −0.1 (−1.9–1.9)	
		June–September	+0.1 (−0.8–0.9) +0.0 (−0.8–0.9)	0.0 (−1.0–1.0) −0.0 (−1.0–0.9)	
	Wave direction change (° clockwise)	December–March	0 (−20–20) 0 (−20–20)	−0 (−20–20) −0 (−20–20)	
		June–September	+0 (−10–10) 0 (−10–10)	0 (−10–10) +0 (−10–10)	
Kiribati (Line Islands)	Wave height change (m)	December–March	−0.0 (−0.3–0.3) −0.0 (−0.3–0.2)	−0.1 (−0.4–0.2) −0.1 (−0.4–0.2)	Low
		June–September	0.0 (−0.2–0.2) +0.0 (−0.1–0.2)	0.0 (−0.2–0.2) +0.0 (−0.2–0.3)	
	Wave period change (s)	December–March	0.0 (−1.5–1.8) −0.1 (−1.4–1.7)	−0.1 (−1.7–1.9) −0.1 (−1.8–2.0)	
		June–September	+0.0 (−0.9–1.0) 0.0 (−0.9–0.9)	0.0 (−1.0–1.0) +0.0 (−1.2–1.0)	
	Wave direction change (° clockwise)	December–March	0 (−30–30) 0 (−20–20)	0 (−30–20) 0 (−30–30)	
		June–September	0 (−10–10) 0 (−10–10)	0 (−10–10) +0 (−10–10)	
Marshall Islands (Northern)	Wave height change (m)	December–March	−0.0 (−0.2 to 0.2) −0.1 (−0.3 to 0.1)	−0.1 (−0.3 to 0.1) −0.2 (−0.5 to 0.0)	Low
		June–September	+0.0 (−0.2 to 0.2) +0.0 (−0.2 to 0.2)	0.0 (−0.2 to 0.2) 0.0 (−0.2 to 0.2)	
	Wave period change (s)	December–March	−0.1 (−0.6 to 0.5) −0.1 (−0.6 to 0.5)	−0.1 (−0.8 to 0.5) −0.2 (−1.0 to 0.5)	
		June–September	−0.0 (−0.9 to 0.8) −0.1 (−0.9 to 0.7)	−0.0 (−0.9 to 0.9) −0.0 (−0.9 to 0.8)	
	Wave direction change (° clockwise)	December–March	0 (−10 to 10) 0 (−10 to 10)	0 (−10 to 10) −0 (−10 to 10)	
		June–September	0 (−10 to 10) +0 (−10 to 20)	+0 (−10 to 20) +10 (−10 to 30)	

continued on next page

Appendix 2 continued

	Variable	Season	2035	2090	Confidence in Projected Changes
Marshall Islands (Southern)	Wave height change (m)	December–March	−0.0 (−0.3 to 0.2) −0.1 (−0.3 to 0.2)	−0.1 (−0.3 to 0.1) −0.2 (−0.4 to 0.0)	Low
		June–September	0.0 (−0.2 to 0.2) +0.0 (−0.2 to 0.2)	0.0 (−0.2 to 0.2) 0.0 (−0.1 to 0.1)	
	Wave period change (s)	December–March	−0.1 (−0.7 to 0.7) −0.1 (−0.7 to 0.8)	−0.1 (−0.8 to 0.8) −0.2 (−1.0 to 0.8)	
		June–September	0.0 (−0.9 to 0.9) −0.1 (−0.9 to 0.8)	0.0 (−0.9 to 0.9) −0.0 (−0.8 to 0.8)	
	Wave direction change (° clockwise)	December–March	0 (−10 to 10) 0 (−10 to 10)	0 (−10 to 10) −0 (−10 to 10)	
		June–September	0 (−20 to 20) 0 (−20 to 20)	+0 (−20 to 30) +10 (−20 to 40)	
Nauru	Wave height change (m)	December–March	−0.0 (−0.2 to 0.2) −0.1 (−0.3 to 0.1)	−0.1 (−0.2 to 0.1) −0.2 (−0.3 to −0.1)	Low
		June–September	+0.0 (−0.1 to 0.1) +0.0 (−0.1 to 0.1)	(−0.1 to 0.1) +0.0 (−0.1 to 0.1)	
	Wave period change (s)	December–March	−0.0 (−1.1 to 1.0) −0.1 (−1.1 to 1.0)	−0.1 (−1.2 to 1.1) −0.2 (−1.3 to 1.0)	
		June–September	+0.0 (−0.6 to 0.7) 0.0 (−0.6 to 0.6)	0.0 (−0.7 to 0.7) −0.1 (−0.8 to 0.6)	
	Wave direction change (° clockwise)	December–March	0 (−10 to 10) 0 (−10 to 10)	0 (−10 to 10) 0 (−10 to 10)	
		June–September	+0 (−10 to 20) +0 (−10 to 20)	+0 (−10 to 20) +10 (−10 to 30)	
Niue	Wave height change (m)	December–March	0.0 (−0.2 to 0.2) −0.0 (−0.2 to 0.2)	−0.0 (−0.3 to 0.2) −0.1 (−0.3 to 0.1)	Low
		June–September	+0.0 (−0.3 to 0.4) +0.0 (−0.3 to 0.4)	0.0 (−0.4 to 0.4) 0.0 (−0.4 to 0.4)	
	Wave period change (s)	December–March	−0.1 (−1.1 to 1.0) −0.1 (−1.1 to 0.9)	−0.1 (−1.2 to 1.1) −0.2 (−1.4 to 1.2)	
		June–September	0.0 (−0.9 to 0.9) 0.0 (−0.9 to 0.9)	0.0 (−1.0 to 1.0) −0.0 (−1.1 to 1.1)	
	Wave direction change (° clockwise)	December–March	0 (−30 to 30) 0 (−30 to 30)	−0 (−30 to 30) −0 (−40 to 30)	
		June–September	−0 (−10 to 10) −0 (−10 to 10)	−0 (−10 to 10) −5 (−10 to 5)	

continued on next page

Appendix 2 continued

	Variable	Season	2035	2090	Confidence in Projected Changes
Palau	Wave height change (m)	December–March	-0.0 (-0.3–0.3) -0.0 (-0.4–0.3)	-0.1 (-0.4–0.2) -0.2 (-0.4–0.1)	Low
		June–September	0.0 (-0.2–0.2) 0.0 (-0.2–0.2)	0.0 (-0.2–0.2) 0.0 (-0.2–0.2)	
	Wave period change (s)	December–March	-0.1(-0.4–0.3) -0.1 (-0.6–0.6)	-0.1 (-0.6–0.4) -0.2 (-0.8–0.4)	
		June–September	-0.0 (-0.6–0.5) -0.0 (-0.6–0.6)	-0.1 (-0.7–0.6) -0.1 (-0.8–0.5)	
	Wave direction change (° clockwise)	December–March	0 (-5–5) 0 (-10–5)	-0 (-10–5) -5 (-10–5)	
		June–September	+0 (-30 to 80) 0 (-0–60)	+0 (-30–80) +10 (-30–70)	
Papua New Guinea	Wave height change (m)	December–March	-0.0 (-0.2–0.1) -0.0 (-0.2–0.1)	-0.1 (-0.2–0.1) -0.1 (-0.2–0.0)	Low
		June–September	0.0 (-0.3–0.3) +0.0 (-0.3–0.3)	+0.0 (-0.2–0.3) +0.0 (-0.2–0.3)	
	Wave period change (s)	December–March	-0.0(-0.9–0.8) -0.1 (-1.0–0.8)	-0.1 (-1.0–0.8) -0.2 (-1.2–0.7)	
		June–September	-0.0 (-0.8–0.7) -0.1 (-0.9–0.8)	-0.1 (-0.9–0.7) -0.2 (-1.1–0.7)	
	Wave direction change (° clockwise)	December–March	0 (-10–10) 0 (-10–10)	-0 (-10–10) -0 (-10–10)	
		June–September	+0 (-40–70) 0 (-40–70)	+0 (-30–60) +10 (-50–70)	
Samoa	Wave height change (m)	December–March	-0.0 (-0.2–0.2) -0.0 (-0.2–0.1)	-0.0 (-0.2–0.1) -0.1 (-0.2–0.2)	Low
		June–September	+0.0 (-0.2–0.3) +0.0 (-0.2–0.2)	(-0.2–0.2) +0.0 (-0.2–0.3)	
	Wave period change (s)	December–March	-0.1(-1.3–1.2) -0.1 (-1.2–1.2)	-0.1 (-1.3–1.5) -0.2 (-1.6–1.6)	
		June–September	0.0 (-0.9–1.0) +0.0 (-1.0–1.0)	+0.0 (-1.1–1.2) 0.0 (-1.3–1.3)	
	Wave direction change (° clockwise)	December–March	0 (-40–40) 0 (-40–30)	0 (-40–40) 0 (-50–50)	
		June–September	0 (-10–10) 0 (-10–10)	0 (-10–10) 0 (-10–10)	

continued on next page

Appendix 2 continued

	Variable	Season	2035	2090	Confidence in Projected Changes
Solomon Islands	Wave height change (m)	December–March	−0.0 (−0.2–0.1) −0.0 (−0.2–0.1)	−0.1 (−0.2–0.1) −0 1 (−0.2–0.0)	Low
		June–September	+0.0 (−0.3–0.3) +0.0 (−0.3–0.3)	+0.0 (−0.3–0.3) +0.0 (−0.3–0.3)	
	Wave period change (s)	December–March	−0.0 (−0.7–0.7) −0.0 (−0.7–0.7)	−0.1 (−1.0–0.7) −0.2 (−1.1–0.8)	
		June–September	−0.0 (−0.6–0.5) −0.0 (−0.6–0.5)	−0.1 (−0.7–0.6) −0.1 (−0.7–0.7)	
	Wave direction change (° clockwise)	December–March	+0 (−30–30) +0 (−30–30)	0 (−30–30) −0 (−30–30)	
		June–September	0 (−5–5) 0 (−5–5)	0 (−5–10) 0 (−5–10)	
Tonga	Wave height change (m)	December–March	0.0 (−0.2–0.2) −0.0 (−0.2–0.2)	−0.0 (−0.3–0.2) −0 1 (−0.3–0.1)	Low
		June–September	+0.0 (−0.3–0.4) +0.0 (−0.3–0.4)	(−0.4–0.4) +0.0 (−0.4–0.4)	
	Wave period change (s)	December–March	−0.1 (−1.0–0.9) −0.1 (−1.0–0.8)	−0.1 (−1.2–0.9) −0.1 (−1.3–1.1)	
		June–September	+0.0 (−0.9–1.0) +0.0 (−0.8–0.9)	+0.0 (−1.1–1.1) 0.0 (−1.2–1.1)	
	Wave direction change (° clockwise)	December–March	−0 (−30–20) 0 (−30–20)	−0 (−30–30) 0 (−30–30)	
		June–September	0 (−10–10) −0 (−10–10)	0 (−10–10) −5 (−10–5)	
Tuvalu	Wave height change (m)	December–March	−0.0 (−0.1–0.1) −0.0 (−0.2–0.1)	−0.1 (−0.2–0.0) −0 1 (−0.2–−0.0)	Low
		June–September	+0.0 (−0.2–0.2) +0.0 (−0.2–0.2)	+0.0 (−0.2–0.3) +0.1 (−0.2–0.3)	
	Wave period change (s)	December–March	−0.0 (−1.1–1.2) −0.1 (−1.2–1.1)	−0.1 (−1.4–1.4) −0.1 (−1.6–1.4)	
		June–September	+0.0 (−1.1–1.1) +0.0 (−1.1–1.4)	+0.0 (−1.3–1.4) 0.0 (−1.4–1.4)	
	Wave direction change (° clockwise)	December–March	+0 (−20–20) 0 (−20–20)	0 (−20–20) +0 (−20–20)	
		June–September	+0 (−10–10) 0 (−10–10)	+0 (−10–10) +0 (−10–10)	

continued on next page

Appendix 2 continued

	Variable	Season	2035	2090	Confidence in Projected Changes
Vanuatu	Wave height change (m)	December–March	−0.0 (−0.2–0.1) −0.0 (−0.2–0.1)	−0.1 (−0.2–0.1) −0 1 (−0.3–0.0)	Low
		June–September	+0.0 (−0.2–0.3) 0.0 (−0.2–0.3)	0.0 (−0.2–0.3) +0.0 (−0.2–0.3)	
	Wave period change (s)	December–March	−0.1 (−0.6–0.4) −0.1 (−0.6–0.5)	−0.1 (−0.7–0.5) −0.2 (−0.8–0.5)	
		June–September	0.0 (−0.5–0.5) 0.0 (−0.5–0.5)	−0.0 (−0.6–0.6) −0.1 (−0.6–0.5)	
	Wave direction change (° clockwise)	December–March	+0 (−10–10) 0 (−10–10)	+0 (−10–10) +0 (−10–10)	
		June–September	0 (−5–5) 0 (−5–5)	0 (−5–10) −0 (−10–5)	

m = meter, RCP = Representative Concentration Pathway, s = second.

Notes: Projected average changes in wave height, period, and direction across Pacific Islands Region nations for December–March and June–September for RCP4.5 (top values) and RCP8.5 (bottom values), for two 20-year periods (2026–2045 and 2081–2100), relative to 1986–2005. The values in brackets represent the 5th to 95th percentile range of uncertainty. Confidence in ranges is expressed as high, medium, or low.

Source: Australian Bureau of Meteorology and Commonwealth Scientific and Industrial Research Organisation (2014).

References

Ashok, K., S.K. Behera, S.A. Rao, H. Weng, and T. Yamagata. 2007. El Niño Modoki and Its Possible Teleconnection. *Journal of Geophysical Research: Oceans.* 112 (C11) doi:10.1029/2006jc003798.

Asian Development Bank (ADB). 2015. *Economic Analysis of Climate-Proofing Investment Projects.* Manila. doi:10.1029/2006jc003798

———. 2020a. *Manual on Climate Change Adjustments for Detailed Engineering Design of Roads Using Examples from Viet Nam.* Manila. http://dx.doi.org/10.22617/TIM200147-2.

Asian Development Bank (ADB). 2020. *Climate Change Adjustments for Detailed Engineering Design of Roads: Experience from Viet Nam.* Manila. http://dx.doi.org/10.22617/TIM200148-2.

Aucan, J. 2018. Effects of Climate Change on Sea Levels and Inundation Relevant to the Pacific Islands. *Science Review.* pp. 43–9. https://reliefweb.int/sites/reliefweb.int/files/.

Australian Bureau of Meteorology and Commonwealth Scientific and Industrial Research Organisation (CSIRO). 2011. Climate Change in the Pacific: Scientific Assessment and New Research. Volume 1: Regional Overview

———. 2014. *Climate Variability, Extremes and Change in the Western Tropical Pacific: New Science and Updated Country Reports 2014.* Melbourne, Australia. http://www.pacificclimatechangescience.org/wp-content/uploads/2014/07/PACCSAP_CountryReports2014_WEB_140710.pdf.

Ayyub, B.M., H.G. Braileanu, and N. Qureshi. 2012. Prediction and Impact of Sea Level Rise on Properties and Infrastructure of Washington, DC. *Risk Analysis.* 32 (11). pp. 1901–18. doi:10.1111/j.1539-6924.2011.01710.x.

Bakker, A.M.R., T.E. Wong, K.L. Ruckert, and K. Keller. 2017. Sea-Level Projections Representing the Deeply Uncertain Contribution of the West Antarctic Ice Sheet. *Scientific Reports.* 7 (1). p. 3880. doi:10.1038/s41598-017-04134-5.

Ballu, V., M.-N. Bouin, P. Siméoni, W.C. Crawford, S. Calmant, J.-M. Boré, T. Kanas, and B. Pelletier. 2011. Comparing the Role of Absolute Sea-Level Rise and Vertical Tectonic Motions in Coastal Flooding, Torres Islands (Vanuatu). *Proceedings of the National Academy of Sciences.* 108 (32). pp. 13019–22. doi:10.1073/pnas.1102842108.

Bamber, J., and R. Riva. 2010. The Sea Level Fingerprint of Recent Ice Mass Fluxes. *The Cryosphere.* 4 (4). pp. 621–27. doi:10.5194/tc-4-621-2010.

Bamber, J.L., M. Oppenheimer, R.E. Kopp, W.P. Aspinall, and R.M. Cooke. 2019. Ice Sheet Contributions to Future Sea-Level Rise from Structured Expert Judgment. *Proceedings of the National Academy of Sciences.* 116 (23). p. 11195. doi:10.1073/pnas.1817205116.

Barnett, J. 2001. Adapting to Climate Change in Pacific Island Countries: The Problem of Uncertainty. *World Development.* 29 (6). pp. 977–93. doi:10.1016/S0305-750X(01)00022-5.

Becker, M., B. Meyssignac, C. Letetrel, W. Llovel, A. Cazenave, and T. Delcroix. 2012. Sea Level Variations at Tropical Pacific Islands since 1950. *Global and Planetary Change.* pp. 80–1, 85–98. doi:10.1016/j.gloplacha.2011.09.004.

Bell, S.S., S.S. Chand, K.J. Tory, A.J. Dowdy, C. Turville, and H. Ye. 2019. Projections of Southern Hemisphere Tropical Cyclone Track Density Using CMIP5 Models. *Climate Dynamics.* 52 (9–10). pp. 6065–79. doi:10.1007/s00382-018-4497-4.

Bindoff, N.L., J. Willebrand, V. Artale, A. Cazenave, J. Gregory, S. Gulev, K. Hanawa, C. le Quéré, S. Levitus, Y. Nojiri, C.K. Shum, L.D. Talley, and A. Unnikrishnan. 2007. Observations: Oceanic Climate Change and Sea Level. In S. Solomon, D. Qin, M. Manning, Z. Chen, M. Marquis, K.B. Averyt, M. Tignor, and H.L. Miller (eds.). *Climate Change 2007: The Physical Science Basis. Contribution of Working Group I to the Fourth Assessment Report of the Intergovernmental Panel on Climate Change.* Cambridge, United Kingdom and New York, NY, USA: Cambridge University Press. pp. 385–432.

Brown, N.J., A. Lal, B. Thomas, S. McClusky, J. Dawson, G. Hu, and M. Jia. 2020. Vertical Motion of Pacific Island Tide Gauges: Combined Analysis from GNSS and Levelling. Record 2020/03. *Geoscience Australia.* Canberra. http://dx.doi.org/10.11636/Record.2020.003.

Brown, P., A. Daigneault, and D. Gawith. 2016. Climate Change and the Economic Impacts of Flooding on Fiji. *Climate and Development.* 5529 (May). pp. 1–12. https://doi.org/10.1080/17565529.2016.1174656.

Cazenave, A. and W. Llovel. 2009. Contemporary Sea Level Rise. *Annual Review of Marine Science.* 2 (1). pp. 145–73. doi:10.1146/annurev-marine-120308-081105.

Church, J.A., P.U. Clark, A. Cazenave, J.M. Gregory, S. Jevrejeva, A. Levermann, M.A. Merrifield, G.A. Milne, R.S. Nerem, P.D. Nunn, A.J. Payne, W.T. Pfeffer, D. Stammer, and A.S. Unnikrishnan. 2013. Sea Level Change. In T.F. Stocker, D. Qin, G.-K. Plattner, M. Tignor, S.K. Allen, J. Boschung, A. Nauels, Y. Xia, V. Bex, and P.M. Midgley (eds.). *Climate Change 2013: The Physical Science Basis. Contribution of Working Group I to the Fifth Assessment Report of the Intergovernmental Panel on Climate Change.* Cambridge, United Kingdom and New York, NY, USA: Cambridge University Press.

Church, J.A., N.J. White, and J.R. Hunter. 2006. Sea-Level Rise at Tropical Pacific and Indian Ocean Islands. *Global and Planetary Change.* 53 (3). pp. 155–68. doi:10.1016/j.gloplacha.2006.04.001.

Connell, J. 2013. *Islands at Risk? Environments, Economies and Contemporary Change.* Edward Elgar Publishing Limited, Cheltenham, United Kingdom.

Daloz, A.S., and S.J. Camargo. 2018. Is the Poleward Migration of Tropical Cyclone Maximum Intensity Associated with a Poleward Migration of Tropical Cyclone Genesis? *Climate Dynamics.* 50 (1). pp. 705–15. doi:10.1007/s00382-017-3636-7.

Deschamps, P., N. Durand, E. Bard, B. Hamelin, G. Camoin, A.L. Thomas, G.M. Henderson, J.I. Okuno, and Y. Yokoyama. 2012. Ice-Sheet Collapse and Sea-Level Rise at the Bølling Warming 14,600 Years Ago. *Nature*. 483 (7391). pp. 559–64. doi:10.1038/nature10902.

Duong, T.M., R. Ranasinghe, A. Luijendijk, D. Walstra, and D. Roelvink. 2017. Assessing Climate Change Impacts on the Stability of Small Tidal Inlets: Part 1—Data Poor Environments. *Marine Geology*. 390. pp. 331–46. doi:https://doi.org/10.1016/j.margeo.2017.05.008.

Duong, T.M., R. Ranasinghe, M. Thatcher, S. Mahanama, Z.B. Wang, P.K. Dissanayake, M. Hemer, A. Luijendijk, J. Bamunawala, D. Roelvink, and D. Walstra. 2018. Assessing Climate Change Impacts on the Stability of Small Tidal Inlets: Part 2—Data Rich Environments. *Marine Geology*. 395. pp. 65–81. doi:https://doi.org/10.1016/j.margeo.2017.09.007.

Elsner, J.B., J.P. Kossin, and T.H. Jagger. 2008. The Increasing Intensity of the Strongest Tropical Cyclones. *Nature*. 455 (7209). pp. 92–5. doi:10.1038/nature07234.

Emanuel, K.A. 2013. Downscaling CMIP5 Climate Models Shows Increased Tropical Cyclone Activity over the 21st Century. *Proceedings of the National Academy of Sciences of the United States of America*. 110 (30). pp. 12219–24. doi:10.1073/pnas.1301293110.

Fairbanks, R.G. 1989. A 17,000-year Glacio-Eustatic Sea Level Record: Influence of Glacial Melting Rates on the Younger Dryas Event and Deep-Ocean Circulation. *Nature*. 342 (6250). pp. 637–42. doi:10.1038/342637a0.

Fox-Kemper, B., H. T. Hewitt, C. Xiao, G. Aðalgeirsdóttir, S.S. Drijfhout, T.L. Edwards, N.R. Golledge, M. Hemer, R.E. Kopp, G. Krinner, A. Mix, D. Notz, S. Nowicki, I.S. Nurhati, L. Ruiz, J.-B. Sallée, A.B.A. Slangen, and Y. Yu. 2021. Ocean, Cryosphere and Sea Level Change. In V. Masson-Delmotte, P. Zhai, A. Pirani, S. L. Connors, C. Péan, S. Berger, N. Caud, Y. Chen, L. Goldfarb, M. I. Gomis, M. Huang, K. Leitzell, E. Lonnoy, J.B.R. Matthews, T. K. Maycock, T. Waterfield, O. Yelekçi, R. Yu, and B. Zhou (eds.). *Climate Change 2021: The Physical Science Basis. Contribution of Working Group I to the Sixth Assessment Report of the Intergovernmental Panel on Climate Change*. Cambridge, United Kingdom: Cambridge University Press.

Garner, A.J., J.L. Weiss, A. Parris, R.E. Kopp, R.M. Horton, J.T. Overpeck, and B.P. Horton. 2018. Evolution of 21st Century Sea Level Rise Projections. *Earth's Future*. 6 (11). pp. 1603–15. doi:10.1029/2018ef000991.

Haigh, I.D., M.D. Pickering, J.A.M. Green, B.K. Arbic, A. Arns, S. Dangendorf, D.F. Hill, K. Horsburgh, T. Howard, D. Idier, D.A. Jay, L. Jänicke, S.B. Lee, M. Müller, M. Schindelegger, S.A. Talke, S.-B. Wilmes, and P.L. Woodworth. 2020. The Tides They Are A-Changin': A Comprehensive Review of Past and Future Nonastronomical Changes in Tides, Their Driving Mechanisms, and Future Implications. *Reviews of Geophysics*. 58 (1). e2018RG000636. doi:https://doi.org/10.1029/2018RG000636.

Hansen, J., M. Sato, P. Hearty, R. Ruedy, M. Kelley, V. Masson-Delmotte, G. Russell, G. Tselioudis, J. Cao, E. Rignot, I. Velicogna, B. Tormey, B. Donovan, E. Kandiano, K. von Schuckmann, P. Kharecha, A.N. Legrande, M. Bauer, and K.W. Lo. 2016. Ice Melt, Sea Level Rise and Superstorms: Evidence from Paleoclimate Data, Climate Modeling, and Modern Observations that 2°C Global Warming Could Be Dangerous. *Atmosphere, Chemistry and Physics*. 16 (6). pp. 3761–812. doi:10.5194/acp-16-3761-2016.

Hemer, M.A., J. Katzfey, and C. Hotan. 2011. *The Wind-Wave Climate of the Pacific Ocean.* Melbourne: The Centre for Australian Weather and Climate Research. https://www.awe.gov.au/sites/default/files/env/pages/275228c5-24db-47f2-bf41-82ef42cda73d/files/wind-wave-report.pdf.

Henderson-Sellers, A., H. Zhang, G. Berz, K. Emanuel, W. Gray, C. Landsea, G. Holland, J. Lighthill, S.-L. Shieh, P. Webster, and K. McGuffie. 1998. Tropical Cyclones and Global Climate Change: A Post-IPCC Assessment. *Bulletin of the American Meteorological Society.* 79 (1). pp. 19–38. doi:10.1175/1520-0477(1998)079%3c0019:Tcagcc%3e2.0.Co;2.

Hinkel, J., J.A. Church, J.M. Gregory, E. Lambert, G. Le Cozannet, J. Lowe, K.L. McInnes, R.J. Nicholls, T.D. van der Pol, and R. van de Wal. 2019. Meeting User Needs for Sea Level Rise Information: A Decision Analysis Perspective. *Earth's Future.* 7 (3). pp. 320–37. doi:https://doi.org/10.1029/2018EF001071.

Hoeke, R.K., K.L. McInnes, J.C. Kruger, R.J. McNaught, J.R. Hunter, and S.G. Smithers. 2013. Widespread Inundation of Pacific Islands Triggered by Distant-Source Wind-Waves. *Global and Planetary Change.* 108. pp. 128–38. doi:10.1016/j.gloplacha.2013.06.006.

Hoeke, R.K., K.L. McInnes, J. O'Grady, F. Lipkin, and F. Colberg. 2014. High Resolution Met-Ocean Modelling for Storm Surge Risk Analysis in Apia, Samoa. *CAWCR Technical Report.* 071. Melbourne: The Centre for Australian Weather and Climate Research. https://www.cawcr.gov.au/technical-reports/CTR_071.pdf.

Hoeke, R.K., K.L. McInnes, and J.G. O'Grady. 2015. Wind and Wave Setup Contributions to Extreme Sea Levels at a Tropical High Island: A Stochastic Cyclone Simulation Study for Apia, Samoa. *Journal of Marine Science and Engineering.* 3 (3). pp. 1117–35.

Hooijer, A. and R. Vernimmen. 2021. Global LiDAR Land Elevation Data Reveal Greatest Sea-Level Rise Vulnerability in the Tropics. *Nature Communications.* 12 (1). p. 3592.

Horton, B.P., N.S. Khan, N. Cahill, J.S.H. Lee, T.A. Shaw, A.J. Garner, A.C. Kemp, S.E. Engelhart, and S. Rahmstorf. 2020. Estimating Global Mean Sea-Level Rise and Its Uncertainties by 2100 and 2300 from an Expert Survey. *npj Climate and Atmospheric Science.* 3 (1). p. 18. doi:10.1038/s41612-020-0121-5.

Horton, B.P., R.E. Kopp, A.J. Garner, C.C. Hay, N.S. Khan, K. Roy, and T.A. Shaw. 2018. Mapping Sea-Level Change in Time, Space, and Probability. *Annual Review of Environment and Resources.* 43 (1). pp. 481–521. doi:10.1146/annurev-environ-102017-025826

Hsu, C.-W. and I. Velicogna. 2017. Detection of Sea Level Fingerprints Derived from GRACE Gravity Data. *Geophysical Research Letters.* 44 (17). pp. 8953–61. doi:10.1002/2017gl074070.

Intergovernmental Panel on Climate Change (IPCC). 2019. *Special Report on the Ocean and Cryosphere in a Changing Climate.* Edited by H.-O. Pörtner, D.C. Roberts, V. Masson-Delmotte, P. Zhai, M. Tignor, E. Poloczanska, K. Mintenbeck, M. Nicolai, A. Okem, J. Petzold, B. Rama, and N. Weyer. Cambridge, UK and New York, USA. https://www.ipcc.ch/srocc/home/.

———. 2021. *Climate Change 2021: The Physical Science Basis. Contribution of Working Group I to the Sixth Assessment Report of the Intergovernmental Panel on Climate Change.* Edited by V. Masson-Delmotte, P. Zhai, A. Pirani, S.L. Connors, C. Péan, S. Berger, N. Caud, Y. Chen, L. Goldfarb, M.I. Gomis, M. Huang, K. Leitzell,

E. Lonnoy, J.B.R. Matthews, T.K. Maycock, T. Waterfield, O. Yelekçi, R. Yu, and B. Zhou. Cambridge, United Kingdom: Cambridge University Press. https://www.ipcc.ch/report/ar6/wg1/.

Kiem, A.S., S.W. Franks, and G. Kuczera. 2003. Multi-Decadal Variability of Flood Risk. *Geophysical Research Letters*. 30 (2). p. 1035. doi:10.1029/2002GL015992.

Knutson, T.R., J.L. McBride, J. Chan, K. Emanuel, G. Holland, C. Landsea, I. Held, J.P. Kossin, A.K. Srivastava, and M. Sugi. 2010. Tropical Cyclones and Climate Change. *Nature Geoscience*. 3 (3). pp. 157–63. doi:10.1038/ngeo779.

Knutson, T., S.J. Camargo, J.C.L. Chan, K. Emanuel, C.-H. Ho, J. Kossin, M. Mohapatra, M. Satoh, M. Sugi, K. Walsh, and L. Wu. 2019. Tropical Cyclones and Climate Change Assessment: Part I: Detection and Attribution. *Bulletin of the American Meteorological Society*. 100 (10). pp. 1987–2007. doi:10.1175/bams-d-18-0189.1.

Knutson, T., S.J. Camargo, J.C.L. Chan, K. Emanuel, C.-H. Ho, J. Kossin, M. Mohapatra, M. Satoh, M. Sugi, K. Walsh, and L. Wu. 2020. Tropical Cyclones and Climate Change Assessment: Part II: Projected Response to Anthropogenic Warming. *Bulletin of the American Meteorological Society*. 101 (3). E303–E322. doi:10.1175/bams-d-18-0194.1.

Kopp, R.E., R.M. DeConto, D.A. Bader, C.C. Hay, R.M. Horton, S. Kulp, M. Oppenheimer, D. Pollard, and B.H. Strauss. 2017. Evolving Understanding of Antarctic Ice-Sheet Physics and Ambiguity in Probabilistic Sea-Level Projections. *Earth's Future*. 5 (12). pp. 1217–33. doi:10.1002/2017ef000663.

Kopp, R.E., R.M. Horton, C.M. Little, J.X. Mitrovica, M. Oppenheimer, D.J. Rasmussen, B.H. Strauss, and C. Tebaldi. 2014. Probabilistic 21st and 22nd Century Sea-Level Projections at a Global Network of Tide-Gauge Sites. *Earth's Future*. 2 (8). pp. 383–406. doi:https://doi.org/10.1002/2014EF000239.

Kuleshov, Y., L. Qi, R. Fawcett, and D. Jones. 2008. On Tropical Cyclone Activity in the Southern Hemisphere: Trends and the ENSO Connection. *Geophysical Research Letters*. 35 (14). L14S08. doi:10.1029/2007GL032983

Kulp, S.A., and B.H. Strauss. 2019. New Elevation Data Triple Estimates of Global Vulnerability to Sea-Level Rise and Coastal Flooding. *Nature Communications*. 10 (1). p. 4844. C:\Users\MO\Documents\ADB\2022 DOC\2022-03-15 PARD\10.1038\s41467-019-12808-z.

Le Bars, D., S. Drijfhout, and H. de Vries. 2017. A High-End Sea Level Rise Probabilistic Projection including Rapid Antarctic Ice Sheet Mass Loss. *Environmental Research Letters*. 12 (4). 044013. doi:10.1088/1748-9326/aa6512.

Magee, A.D., and D.C. Verdon-Kidd. 2019. Historical Variability of Southwest Pacific Tropical Cyclone Counts Since 1855. *Geophysical Research Letters*. 46 (12). pp. 6936–45. doi:10.1029/2019gl082900.

Magee, A.D., D.C. Verdon-Kidd, H.J. Diamond, and A.S. Kiem. 2017. Influence of ENSO, ENSO Modoki, and the IPO on Tropical Cyclogenesis: A Spatial Analysis of the Southwest Pacific Region. *International Journal of Climatology*. 37. pp. 1118–37. doi:10.1002/joc.5070.

Masselink, G., E. Beetham, and P. Kench. 2020. Coral Reef Islands Can Accrete Vertically in Response to Sea Level Rise. *Science Advances.* 6 (24). eaay3656. doi:doi:10.1126/sciadv.aay3656.

McInnes, K.L., R.K. Hoeke, K.J.E. Walsh, J.G. O'Grady, and G.D. Hubbert. 2016a. Application of a Synthetic Cyclone Method for Assessment of Tropical Cyclone Storm Tides in Samoa. *Natural Hazards.* 80 (1). pp. 425–44. doi:10.1007/s11069-015-1975-4.

McInnes, K.L., J.G. O'Grady, K.J.E. Walsh, and F. Colberg. 2011. Progress Towards Quantifying Storm Surge Risk in Fiji due to Climate Variability and Change. *Journal of Coastal Research.* pp. 1121–4.

McInnes, K.L., K.J.E. Walsh, R.K. Hoeke, J.G. O'Grady, F. Colberg, and G.D. Hubbert. 2014. Quantifying Storm Tide Risk in Fiji due to Climate Variability and Change. *Global and Planetary Change.* 116. pp. 115–29. doi:https://doi.org/10.1016/j.gloplacha.2014.02.004.

McInnes, K.L., C.J. White, I.D. Haigh, M.A. Hemer, R.K. Hoeke, N.J. Holbrook, A.S. Kiem, E.C.J. Oliver, R. Ranasinghe, K.J.E. Walsh, S. Westra, and R. Cox. 2016b. Natural Hazards in Australia: Sea Level and Coastal Extremes. *Climatic Change.* 139 (1). pp. 69–83. doi:10.1007/s10584-016-1647-8.

Merrifield, M.A., Y. Firing, and J. Marra. 2007. Annual Climatologies of Extreme Water Levels. *Proc. Aha Hulikoa: Extreme Events—Hawaiian Winter Workshop.* Manoa, HI, USA: University of Hawaii. 23–26 January.

Moftakhari, H.R., G. Salvadori, A. AghaKouchak, B.F. Sanders, and R.A. Matthew. 2017. Compounding Effects of Sea Level Rise and Fluvial Flooding. *Proceedings of the National Academy of Sciences.* 201620325. doi:10.1073/pnas.1620325114

Mori, N., T. Shimura, K. Yoshida, R. Mizuta, Y. Okada, M. Fujita, T. Khujanazarov, and E. Nakakita. 2019. Future Changes in Extreme Storm Surges Based on Mega-Ensemble Projection Using 60-Km Resolution Atmospheric Global Circulation Model. *Coastal Engineering Journal.* 61 (3). pp. 295–307. doi:10.1080/21664250.2019.1586290.

Morim, J., M. Hemer, X.L. Wang, N. Cartwright, C. Trenham, A. Semedo, I. Young, L. Bricheno, P. Camus, M. Casas-Prat, L. Erikson, L. Mentaschi, N. Mori, T. Shimura, B. Timmermans, O. Aarnes, Ø. Breivik, A. Behrens, M. Dobrynin, M. Menendez, J. Staneva, M. Wehner, J. Wolf, B. Kamranzad, A. Webb, J. Stopa, and F. Andutta. 2019. Robustness and Uncertainties in Global Multivariate Wind-Wave Climate Projections. *Nature Climate Change.* 9 (9). pp. 711–18. doi:10.1038/s41558-019-0542-5.

Morim, J., C. Trenham, M. Hemer, X.L. Wang, N. Mori, M. Casas-Prat, A. Semedo, T. Shimura, B. Timmermans, P. Camus, L. Bricheno, L. Mentaschi, M. Dobrynin, Y. Feng, and L. Erikson. 2020. A Global Ensemble of Ocean Wave Climate Projections from CMIP5-Driven Models. *Scientific Data.* 7 (1). p. 105. doi:10.1038/s41597-020-0446-2.

Muis, S., M.I. Apecechea, J. Dullaart, J. de Lima Rego, K.S. Madsen, J. Su, K. Yan, and M. Verlaan. 2020. A High-Resolution Global Dataset of Extreme Sea Levels, Tides, and Storm Surges, including Future Projections. *Frontiers in Marine Science.* 7 (263). doi:10.3389/fmars.2020.00263.

National Climate Change Adaptation Research Facility (NCCARF). 2016. Guidance on Undertaking a First-Pass Risk Screening. CoastAdapt, NCCARF, Gold Coast, Australia. https://coastadapt.com.au/sites/default/files/factsheets/T3M4_1_1st_pass_risk_assessment_0.pdf.

National Oceanic and Atmospheric Administration (NOAA). 2017. Global and Regional Sea Level Rise Scenarios for the United States. *NOAA Technical Report.* NOS CO-OPS 083. Silver Spring, MD, USA. https://tidesandcurrents.noaa.gov/publications/techrpt83_Global_and_Regional_SLR_Scenarios_for_the_US_final.pdf.

Needham, H.F., B.D. Keim, and D. Sathiaraj. 2015. A Review of Tropical Cyclone-Generated Storm Surges: Global Data Sources, Observations, and Impacts. *Reviews of Geophysics.* 53 (2). pp. 545–91. doi:10.1002/2014RG000477.

Nerem, R.S., D.P. Chambers, C. Choe, and G.T. Mitchum. 2010. Estimating Mean Sea Level Change from the TOPEX and Jason Altimeter Missions. *Marine Geodesy.* 33. pp. 435–46. doi:10.1080/01490419.2010.491031.

Nicholls, R.J., S.E. Hanson, J.A. Lowe, A.B.A. Slangen, T. Wahl, J. Hinkel, and A.J. Long. 2021. Integrating New Sea-Level Scenarios into Coastal Risk and Adaptation Assessments: An Ongoing Process. *WIREs Climate Change.* 12 (3). e706. doi:https://doi.org/10.1002/wcc.706.

Nicholls, R.J., S.E. Hanson, J.A. Lowe, R.A. Warrick, X. Lu, and A.J. Long. 2014. Sea-Level Scenarios for Evaluating Coastal Impacts. *WIREs Climate Change.* 5 (1). pp. 129–50. doi:https://doi.org/10.1002/wcc.253.

Nurse, L.A., R.F. McLean, J. Agard, L.P. Briguglio, V. Duvat-Magnan, N. Pelesikoti, E. Tompkins, and A. Webb. 2014. Small Islands. In V.R. Barros, C.B. Field, D.J. Dokken, M.D. Mastrandrea, K.J. Mach, T.E. Bilir, M. Chatterjee, K.L. Ebi, Y.O. Estrada, R.C. Genova, B. Girma, E.S. Kissel, A.N. Levy, S. MacCracken, P.R. Mastrandrea, and L.L. White (eds.). *Climate Change 2014: Impacts, Adaptation, and Vulnerability. Part B: Regional Aspects. Contribution of Working Group II to the Fifth Assessment Report of the Intergovernmental Panel on Climate Change.* Cambridge, United Kingdom and New York, NY, USA: Cambridge University Press. pp. 1613–54.

Oliver, E.C.J., and K.R. Thompson. 2010. Madden-Julian Oscillation and Sea Level: Local and Remote Forcing. *Journal of Geophysical Research: Oceans.* 115 (1). pp. 1–15. doi:10.1029/2009JC005337.

Pacific Region Infrastructure Facility (PRIF). 2021. Guidance for Managing Sea Level Rise Infrastructure Risk in Pacific Island Countries.

Parker, C.L., C.L. Bruyère, P.A. Mooney, and A.H. Lynch. 2018. The Response of Land-Falling Tropical Cyclone Characteristics to Projected Climate Change in Northeast Australia. *Climate Dynamics.* 51 (9). pp. 3467–85. doi:10.1007/s00382-018-4091-9

Patricola, C.M., and M.F. Wehner. 2018. Anthropogenic Influences on Major Tropical Cyclone Events. *Nature.* 563 (7731). pp. 339–46. doi:10.1038/s41586-018-0673-2.

Pickering, M.D., K.J. Horsburgh, J.R. Blundell, J.J.M. Hirschi, R.J. Nicholls, M. Verlaan, and N.C. Wells. 2017. The Impact of Future Sea-Level Rise on the Global Tides. *Continental Shelf Research.* 142. pp. 50–68. doi:https://doi.org/10.1016/j.csr.2017.02.004.

Quataert, E., C. Storlazzi, A. van Rooijen, O. Cheriton, and A. van Dongeren. 2015. The Influence of Coral Reefs and Climate Change on Wave-Driven Flooding of Tropical Coastlines. *Geophysical Research Letters.* 42 (15). pp. 6407–15. doi:https://doi.org/10.1002/2015GL064861.

Ramsay, D. 2011. Synthesis report prepared for the Department of Climate Change and Energy Efficiency, Government of Australia..

Reguero, B.G., I.J. Losada, and F.J. Méndez. 2019. A Recent Increase in Global Wave Power as a Consequence of Oceanic Warming. *Nature Communications.* 10 (1). p. 205. doi:10.1038/s41467-018-08066-0.

Sharmila, S., and K.J.E. Walsh. 2018. Recent Poleward Shift of Tropical Cyclone Formation Linked to Hadley Cell Expansion. *Nature Climate Change.* 8 (8). pp. 730–6. doi:10.1038/s41558-018-0227-5.

Slater, T., I.R. Lawrence, I.N. Otosaka, A. Shepherd, N. Gourmelen, L. Jakob, P. Tepes, L. Gilbert, and P. Nienow. 2021. Review Article: Earth's Ice Imbalance. *The Cryosphere.* 15 (1). pp. 233–46. doi:10.5194/tc-15-233-2021.

Solomon, S.M., and D.L. Forbes. 1999. Coastal Hazards and Associated Management Issues on South Pacific Islands. *Ocean and Coastal Management.* 42 (6–7). pp. 523–54. doi:10.1016/S0964-5691(99)00029-0.

Sonel. Vertical Land Movements. https://www.sonel.org/-Vertical-land-movement-estimate-.html?lang=en.

Stammer, D., R.S.W. van de Wal, R.J. Nicholls, J.A. Church, G. Le Cozannet, J.A. Lowe, B.P. Horton, K. White, D. Behar, and J. Hinkel. 2019. Framework for High-End Estimates of Sea Level Rise for Stakeholder Applications. *Earth's Future.* 7 (8). pp. 923–38. doi:https://doi.org/10.1029/2019EF001163.

Stephens, S.A., R.G. Bell, D. Ramsay, and N. Goodhue. 2014. High-Water Alerts from Coinciding High Astronomical Tide and High Mean Sea Level Anomaly in the Pacific Islands Region. *Journal of Atmospheric and Oceanic Technology.* 31 (12). pp. 2829–43. doi:10.1175/JTECH-D-14-00027.1.

Sterl, A. 2004. On the (In)Homogeneity of Reanalysis Products. *J. Climate.* 17. pp. 3866–73.

Storlazzi, C.D., S.B. Gingerich, A.V. Dongeren, O.M. Cheriton, P.W. Swarzenski, E. Quataert, C.I. Voss, D.W. Field, H. Annamalai, G.A. Piniak, and R. McCall. 2018. Most Atolls Will Be Uninhabitable by the Mid-21st Century Because of Sea-Level Rise Exacerbating Wave-Driven Flooding. *Science Advances.* 4 (4). eaap9741. doi:doi:10.1126/sciadv.aap9741.

Sugi, M., K. Yoshida, and H. Murakami. 2015. More Tropical Cyclones in a Cooler Climate? *Geophysical Research Letters.* 42 (16). pp. 6780–4. doi:10.1002/2015gl064929.

Sweet, W.V., R. Horton, R.E. Kopp, A.N. LeGrande, and A. Romanou. 2017. Sea Level Rise. In D.J. Wuebbles, D.W. Fahey, K.A. Hibbard, D.J. Dokken, B.C. Stewart, and T.K. Maycock (eds.). *Climate Science Special Report: Fourth National Climate Assessment, Volume I.* Washington, DC: US Global Change Research Program. pp. 333–63. doi:%2010.7930/J0VM49F2.

Terry, J.P., S. McGree, and R. Raj. 2004. The Exceptional Flooding on Vanua Levu Island, Fiji, during Tropical Cyclone Ami in January 2003. *Journal of Natural Disaster Science.* 26 (1). pp. 27–36.

Thorne, P. 2015. Interactive Comment on "Ice Melt, Sea Level Rise and Superstorms: Evidence From Paleoclimate Data, Climate Modeling, and Modern Observations that 2°C Global Warming Is Highly Dangerous" by J. Hansen et al. *Atmos. Chem. Phys. Discuss.* 15. C6089–C6100. https://acp.copernicus.org/preprints/15/C6089/2015/acpd-15-C6089-2015.pdf.

Timmermans, B.W., C.P. Gommenginger, G. Dodet, and J.-R. Bidlot. 2020. Global Wave Height Trends and Variability from New Multimission Satellite Altimeter Products, Reanalyses, and Wave Buoys. *Geophysical Research Letters.* 47 (9). e2019GL086880. doi:https://doi.org/10.1029/2019GL086880.

Trenham, C.E., M.A. Hemer, T.H. Durrant, and D.J.M. Greenslade. 2013. PACCSAP Wind-Wave Climate: High Resolution Wind-Wave Climate and Projections of Change in the Pacific Region for Coastal Hazard Assessments. *CAWCR Technical Report.* 068. Melbourne: The Centre for Australian Weather and Climate Research. https://www.cawcr.gov.au/technical-reports/CTR_068.pdf.

United States National Aeronautics and Space Administration (NASA) and Intergovernmental Panel on Climate Change. Sea Level Projection Tool. https://sealevel.nasa.gov/ipcc-ar6-sea-level-projection-tool.

Vitousek, S., P.L. Barnard, C.H. Fletcher, N. Frazer, L. Erikson, and C.D. Storlazzi. 2017. Doubling of Coastal Flooding Frequency within Decades due to Sea-Level Rise. *Scientific Reports.* 7 (1). p. 1399. doi:10.1038/s41598-017-01362-7.

Walsh, K. 2015. Fine Resolution Simulations of the Effect of Climate Change on Tropical Cyclones in the South Pacific. *Climate Dynamics.* 45 (9–10). pp. 2619–31. doi:10.1007/s00382-015-2497-1.

Walsh, K.J.E., J.L. McBride, P.J. Klotzbach, S. Balachandran, S.J. Camargo, G. Holland, T.R. Knutson, J.P. Kossin, T.-c. Lee, A. Sobel, and M. Sugi. 2016. Tropical Cyclones and Climate Change. *Wiley Interdisciplinary Reviews: Climate Change.* 7 (1). pp. 65–89. doi:10.1002/wcc.371.

Walsh, K.J.E., K.L. McInnes, and J.L. McBride. 2012. Climate Change Impacts on Tropical Cyclones and Extreme Sea Levels in the South Pacific—A Regional Assessment. *Global and Planetary Change.* 80–81, pp. 149–64. C:\Users\MO\Documents\ADB\2022 DOC\2022-03-15 PARD\10.1016\j.gloplacha.2011.10.006

Wandres, M., J. Aucan, A. Espejo, N. Jackson, A. de Ramon N'Yeurt, and H. Damlamian. 2020. Distant-Source Swells Cause Coastal Inundation on Fiji's Coral Coast. *Frontiers in Marine Science.* 7 (546). doi:10.3389/fmars.2020.00546.

Wang, J., J.A. Church, X. Zhang, and X. Chen. 2021. Reconciling Global Mean and Regional Sea Level Change in Projections and Observations. *Nature Communications.* 12 (1). p. 990. doi:10.1038/s41467-021-21265-6.

White, N.J., I.D. Haigh, J.A. Church, T. Koen, C.S. Watson, T.R. Pritchard, P.J. Watson, R.J. Burgette, K.L. McInnes, Z.-J. You, X. Zhang, and P. Tregoning. 2014. Australian Sea Levels—Trends, Regional Variability and Influencing Factors. *Earth-Science Reviews.* 136. pp. 155–74. doi:https://doi.org/10.1016/j.earscirev.2014.05.011.

Wilby, R.L., R.J. Nicholls, R. Warren, H.S. Wheater, D. Clarke, and R.J. Dawson. 2011. Keeping Nuclear and Other Coastal Sites Safe from Climate Change. *Proceedings of the Institution of Civil Engineers—Civil Engineering.* 164 (3). pp. 129–36. doi:10.1680/cien.2011.164.3.129\.

Wilby, R., R. Dawson, C. Kilsby, R.J. Nicholls, and R. Warren. 2014. How to Define Credible Maximum Sea-Level Change Scenarios for the UK Coast. *British Energy Estuarine and Marine Studies (BEEMS) Scientific Advisory Report.* Series 2014, no. 024.United Kingdom: British Energy Climate Change Working Group.

Winter, G., C. Storlazzi, S. Vitousek, A. van Dongeren, R. McCall, R. Hoeke, W. Skirving, J. Marra, J. Reyns, J. Aucan, M. Widlansky, J. Becker, C. Perry, G. Masselink, R. Lowe, M. Ford, A. Pomeroy, F. Mendez, A. Rueda, and M. Wandres. 2020. Steps to Develop Early Warning Systems and Future Scenarios of Storm Wave-Driven Flooding Along Coral Reef-Lined Coasts. *Frontiers in Marine Science.* 7 (199). doi:10.3389/fmars.2020.00199.

Wong, T.E., A.M.R. Bakker, and K. Keller. 2017. Impacts of Antarctic Fast Dynamics on Sea-Level Projections and Coastal Flood Defense. *Climatic Change.* 144 (2). pp. 347–64. doi:10.1007/s10584-017-2039-4.

Young, I.R., and A. Ribal. 2019. Multiplatform Evaluation of Global Trends in Wind Speed and Wave Height. *Science.* 364 (6440). pp. 548–52. doi:10.1126/science.aav9527.

Zhang, X., and J.A. Church. 2012. Sea Level Trends, Interannual and Decadal Variability in the Pacific Ocean. *Geophysical Research Letters.* 39 (21). pp. 1–8. doi:10.1029/2012GL053240.

Lightning Source UK Ltd.
Milton Keynes UK
UKHW052218251022
411101UK00014B/81